# COMPUTERIZATION
## IN THE
## WATER AND
## WASTEWATER FIELDS

Edited by

Eugene A. Glysson
Eric J. Way
Richard W. Force
Wayne H. Abbott, Jr.

LEWIS PUBLISHERS, INC.

**Library of Congress Cataloging-in-Publication Data**

Computerization in the water and wastewater fields.

Includes index.
1. Water-supply — Data processing.   2. Sewage disposal
plants — Data processing.   3. Microcomputers.   I. Glysson,
E. A.
TD353.C63      1986      628.1'028'5      86-20146
ISBN 0-87371-025-8

LEWIS PUBLISHERS, INC.
121 South Main Street, P.O. Drawer 519, Chelsea, Michigan  48118

PRINTED IN THE UNITED STATES OF AMERICA

# PREFACE

There are many ways that a microcomputer may be useful in the water and wastewater fields. This book is intended to provide information leading to a better understanding of the computer itself and to show how it can be effectively and efficiently used in both large and small plants. The book is intended to be of interest to anyone who wishes to employ this modern data handling device.

Contents of the text include a discussion of the selection of microcomputer and its software from a user's standpoint. Once you have clearly determined what you want the computer to do, you can proceed to acquire the software that can accomplish that objective and the computer to utilize that software. The prospective user must give the computer application considerable thought in advance to assure that what is selected will be compatible with the personnel who use it and the information to be processed.

Microcomputers are finding an increasing role in the operation of water and wastewater plants. Their typical initial uses are in word processing, report writing, correspondence, inventory control, and general bookkeeping functions. However, they can be used for acquiring signals, monitoring certain conditions in a plant, and taking action based on the signals received. This book includes the application of very simple and basic examples of transmitting input signals to a variety of low-cost microcomputers. These applications can be utilized by both water and wastewater facilities of both large and small capacities.

There are many other uses of the microcomputer in the water and wastewater fields addressed in this volume. Included are:

- Utility rate studies
- Water and sewer network analysis
- Mapping and design

The optimization of power utilization lends itself to computer analysis and control, resulting in a reduction of energy consumption at both large and small plants. Computer methods of revising plant operations to achieve an optimum utility bill are discussed.

Maintenance organization and implementation can be greatly improved by computer applications. This improvement in operations is discussed as utilized in a large metropolitan installation.

It is the goal of the authors and editors of this book that the microcomputer become the useful tool that it was intended to be and that it be more widely utilized in the water and wastewater fields.

Eugene A. Glysson
Eric J. Way
Richard W. Force
Wayne H. Abbott, Jr.

# CONTENTS

# CHAPTER 1

# AN INTRODUCTION TO COMPUTERS

Brice Carnahan, Ph.D.
Department of Chemical Engineering
University of Michigan
Ann Arbor, Michigan

As edited by
Eugene A. Glysson, P.E., Ph.D.
Department of Civil Engineering
University of Michigan
Ann Arbor, Michigan

## INTRODUCTION

Many readers may have computing experience and some may be very knowledgeable about computers. However, this chapter assumes that the reader knows nothing about them and will start from scratch. It will try to bring you up-to-date on what is going on, especially in the microcomputer area, because that is probably the part of the technology that you are most likely to be dealing with in municipal waterworks and related areas. The following topics will be discussed:

> Machine Organization - how computers are organized from a functional standpoint. We will discuss a little bit about hardware characteristics, i.e., information abut speeds and capacities - the things you would want to know as a user if you were going to buy a machine.

> Operating Systems - special programs that allow you to use the computer effectively.

> Programming Languages - which allow you to write orders for the computer. Not everyone will be doing this because in many cases you will be able to buy the programs that you need or have others write them for you.

> System Programs - programs that are generally useful for managing the computing system and, in some cases, managing the office.

> Application Programs - programs that solve problems of interest to the users.

Networks and Communication - linking computers together, communication between computers.

Let us begin by discussing the kinds of computers that we are going to deal with. First of all, they are called digital computers. There are other kinds of electronic machines, but basically when someone says computer these days, they mean digital computer. The machines that we use are called general purpose digital computers. Computer scientists have a definition of what computability is. Essentially, computable problems are those for which one can write procedures which are essentially unambiguous sequences of orders or instructions for solving problems. These procedures have to be written in some sort of symbolic form or mathematical notation, but they do have some special characteristics that make them possible to implement on machines. For one thing, they have to be unambiguous procedures. That is, at every stage it must be clear what is to be done next. Secondly, the procedures must be finite. That is, they have to be terminating. We cannot evaluate an infinite series on a computer, but we may be able to evaluate the most important part - the first few terms in a series, for example, and the computer probably can do that quite adequately. Algorithm then is probably the preferred word by computerniks for procedures. Flow diagrams are simply graphical representations of procedures the computer follows, and if we have a procedure in the form the computer could really do something with directly, then we call it a program.

## MACHINE ORGANIZATION

Basically, computing machines are very much like most of the other equipment that we use these days. We can use them pretty effectively without knowing what really makes them function. We can drive cars without knowing anything about internal combustion engines. We can use television sets without knowing anything about cathode ray tubes. We do have to know a few things as a computer user. First of all, what can the machine do for us? Secondly, how do we communicate our intentions to the machine so that it will do what it is supposed to be able to do for us? Thirdly, are we sure that the machine is going to do what we have told it to do? Fortunately, the latter question, which was a problem in the early days of computers, is a moot point now. Most machines are extremely reliable. They are not infallible. Every machine has the possibility of breaking or not functioning, but the fact is that computing machines are among the most reliable pieces of equipment that we can put together. It is not unlikely the machine could run for hours and hours and hours doing millions of calculations per second without making a single digit error. There might be internal errors, but in many cases they are caught and corrected, and as far as the user is concerned there would be no evidence that any error had occurred.

## Input/Output Hardware

When it comes to putting a machine together, you really need to look at the functional details rather than the hardware details. Almost all computers are organized around some basic functional elements which are implemented in hardware. Somehow we have to be able to communicate with the machine, and the machine has to be able to get back to us with results of all its efforts. We call that the input/output equipment, or the I/O equipment. Typically, we have a keyboard for input and a screen for ouput. We could attach a printer for output. The disk drives are also input/output devices. But I/O functions alone would not enable us to do much. We need some computational capability.

## Processing Units

The part of the machine that does the computation is usually called the arithmetic unit. Sometimes it's called the ALU, the arithmetic and logic unit, because machines do more than just arithmetic. In order to make the whole thing function and in fact carry out a particular sequence of instructions, there must be a part of the machine that tells the rest of the machine what to do and when to do it. That part of the machine are called a control unit. These two units taken together, the arithmetic unit and the control unit, called the central processing unit (CPU). In the new microcomputers, the two units are fabricated together on a single chip called a microprocessor; that's why it's called a microcomputer.

Computers are identified by their manufacturer and a model number. For instance, here at The University of Michigan, we have Amdol equipment. We would say we have an Amdol 5860. Anybody who is in the computing business would know exactly what that meant in terms of the arithmetic and control unit. That, however, would not tell anyone very much about the other parts of the machine, because the other parts tend to be like accessories or options. We could have different amounts of memory or different speeds of memory, different kinds of input and output equipment attached, and so forth. The basic processor itself, when we identify it by number and manufacturer, would be fairly clear. If you went to a different machine with the same number you would find the same processor, but perhaps not the same memory and/or I/O equipment. In small machines, you usually do not care much about the processor itself. We hear about the IBM Personal Computer (PC) but there are really several different ones. In fact there are different processors in different versions of the IBM PC equipment.

## Memory

What about memory? When people talking about a computer use the term "memory," they are talking about the main central or fast memory. This is the part of the memory to which the processor has direct access. It is very high speed, completely electronically accessible, with no moving parts. The characteristic of this memory is that we can read information from the memory without destroying it. It is like a tape recorder: we can play the song over and over and we get the same piece of information when we ask for it. On the other hand, if we put a new piece of information into a memory element, then we destroy what was there previously. This is because the elements that do the storing are physical, such as a small magnet or perhaps a small capacitor, and are used to represent information that is numerical or non-numerical. If we change the configuration of the magnets or modify which capacitors are charged and which are not, we then destroy any information which was represented there before by entering the new set of charges. This memory - the main central or fast memory - can be accessed very rapidly. Even on very inexpensive computers, the typical access time would be on the order of microseconds. On a very large machine, the memory would be accessible at least one hundred, and may be even one thousand times faster. These are access times in nanoseconds - billionths of a second. This memory is called random access memory (RAM) because it doesn't make any difference which part of the memory you try to access: it takes exactly the same amount of time to retrieve the information. In that sense, it is quite different than some of our nonrandom access memories, such as magnetic tapes and disk drives, where the position of the device determines how long it takes to get a piece of information.

When you buy a machine, especially a small machine, you will usually hear someone say, "I've got an IBM PC with 256 K RAM." That means they have 256 kilobytes of this very fast memory. No information can actually be processed by the computer without its being resident in main memory. The memory is really the heart of the machine where information is stored and modified. How much of this memory do we get? On a small machine, such as the smallest IBM PC that you could buy, we would have 64,000 bytes. A byte is a sequence of individual storage elements which can each be bi-status; that is they can store one of two values. Inside our computers, all numbers and characters are represented as strings of zeros and ones, each letter and number having a unique pattern. Accordingly, we cannot use standard decimal base arithmetic inside the computer. Instead, information is stored and manipulated in binary form, meaning all the information is represented as either a zero or a one. Each one of these little bi-status devices, when represented as part of the memory, is called a bit. Typically, in the memories that are organized along IBM lines, one byte contains eight of these bits. A byte can store either a number, a letter, or a coded instruction to the machine. When we say memory has 512 K bytes, then inside the machine's memory we have 512,000 times 8 of these little bits of usable information. Actually there are some additional bits that check for simple parity. This is a way of keeping errors under control. As far as the user is concerned, we can think of memory as being 8 times roughly 500,000, which means we have a very large number of these small memory elements even in a small machine.

Since we have so many memory elements, somehow we have to be able to identify the first byte from the second from the third in the 510,000 and so forth. We do this usually by giving an address to each of the bytes. We take our collection of all these little storage elements, and we group them together individually in groups of eight - the bytes. We then give each one of these bytes an address in the memory so that we can refer to it when we want to save something or to retreive it. The first byte is typically given the address 000 and the next one 001 and the next one 002, but of course it wouldn't be 002 in binary code, it would be 010. Even the addresses themselves are in terms of binary information. Everything about the machine inside is binary in nature because the elements that make up all the circuits and the memory and so forth are all binary. They are essentially switches which are on or off, capacitors that are charged or not charged, or magnets with different polarity. We don't really have to worry about the binary language because all machines that are now being sold have interfaces for the users so you can deal with the computer in characters that you are used to using, i.e., the Roman alphabet, punctuation marks and standard decimal notation. But inside the machine, that information has to be transformed somehow into binary form.

A major difference between microcomputers and large computers is the number of bytes that can be dealt with at the same time. For example, an Apple II is called an 8 bit machine because it deals with information one byte at a time. An IBM PC has a different microprocessor unit which can deal with two bytes at a time. It is called 16 bit microprocessor. Other microcomputers can deal with 4 bytes at a time, and are called 32 bit microprocessors. Super computers, very large machines, can deal with even larger numbers of bytes at one time. The categorization of computers as to size, speed and so forth, usually has something to do with how fast individual bytes of information can be accessed and manipulated, and how many of them can be dealt with at the same time.

## Auxiliary Storage

High speed memory, although it's getting cheaper all the time, is still fairly expensive. Even though 250,000 bytes sounds like a lot of space, when filled with 250,000 characters it is not very much text. So we cannot keep all the information that we really might want to keep in the machine or have associated with the machine. We need some other storage devices. Currently, the standard storage devices that are used are magnetic tapes and magnetic disks. There is a whole new area of technology opening up and it won't be long before we will be using laser disks. Compact disk audio technology is already here and something very similar should soon be available to microcomputer users. Video disks as well will probably be available as storage devices.

## Magnetic Tape

There are going to be all kinds of other secondary storage equipment available for machines in the future; but at the moment the typical large machine has magnetic tape drives, while microcomputers do not. Magnetic tape is very cheap. The tape is 1/2 inch wide and the bytes are written across the tape in tracks; the standard tape drives have 9 tracks. Eight of those tracks have to do with the 8 bits in a byte. High-density tapes have over 6,000 of these bytes written in one inch of track. Therefore, in one inch of a 1/2-inch tape, there are about 6,000 times 9 or 50,000 little magnets being written on. Some of our tapes now can read or write at the rate of 150 inches per second. We can access information very rapidly from magnetic tape. Just multiply those numbers together and you can see this is a lot of bytes - something on the order of a megabyte per second. Unfortunately, since the tape is wound, it might happen that the information needed is way on the inside of the spool. In this case, it could take several seconds to spin the tape to get it to the point where the information needed can be accessed.

The information on the tape has to be recorded in a way that will allow determination of where on the tape the particular piece of information lies. To do this, the different bytes are put together into records. The records are further put together into files. There are special marks in the tape to indicate where the end of each record and the end of each file occur, so if one knows that a piece of desired information is in the 4th record of the 7th file on the tape, the machine can be instructed to spin the tape to that point and then start reading the information. Obviously, this is not a random access device, as RAM memory is, because it takes a varying amount of time to get to the information and access it, depending on the position of the tape. Once the tape is in the proper position, the data acquisition rate can be quite high. How the information is organized on the tape is important. All the information you think you are going to use at one time should be close together on the tape so that a lot of time is not spent moving around from one place to another just to pick up one number here and one number there. Data organization on nonrandom access storage devices is something of an art: one has to be pretty careful about how it is done.

## Magnetic Disks

The important technology from the standpoint of microcomputers is disk storage. Disk storage essentially involves a magnetic rotating storage medium - the floppy disk, or diskette (Figure 1). Typically the floppy disk is about 5-1/4 inches in diameter. It is simply a piece of mylar which has a magnetizable surface on it, very much like the surface of a magnetic tape. It is stored in a paper sleeve. There is a lubricant between the sleeve and the storage medium so that it spins inside the sleeve. When you insert the sleeve into one of the disk drives, it will clamp around the center mark of the storage media and then start spinning it. On most microcomputers, the spin rate is about 300 rpm, which is not very fast. The information is written in circular tracks on the surface of the disk. The most commonly used disks have 40 tracks on each side. Each track is divided into 9 pie-shaped pieces, call sectors. Each of the sectors has 512 bytes, and each byte has 8 bits, so with 40 tracks time 9 sectors there are a lot of bits being written on the surface of the diskette. When put together, we find there are 184,000 bytes on each side of the diskette. If we multiply that times two, we get 368,000 bytes on one floppy disk - called a 360K diskette.

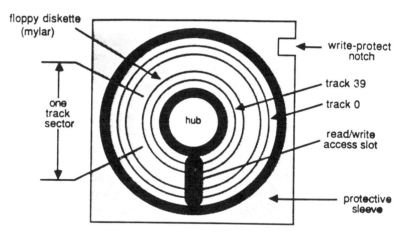

**Figure 1.** Parts of a diskette.

This sounds like a lot of space to store information on, but in fact it really isn't. It takes no time at all to fill up one of these diskettes. Simply writing a report may take a few thousand bytes. Write another report and it takes up a few thousand more bytes. After two or three weeks of using the computer you will have one diskette full. Since it is very easy to generate lots of information, we generally need another storage device which has a bigger capacity. The latest of the available IBM PCs is called the AT. The AT has a floppy disk drive which takes the same size diskettes but they can hold 1.2 megabytes instead of 360 kilobytes. This is due to having a better mechanism for aligning the heads so that the bits can be written more closely together. The technology keeps improving. We will see diskettes that will hold twice that much within the next year or two. Even though there seems to be almost no limit to the cleverness of the people who are working on storage devices, we are probably not going to be able to keep 100 megabytes on a floppy diskette. The mechanics of the drive system are simply not precise enough.

The limitations of floppy disks can be surmounted by using hard disks. These are disks that have rigid metal plates. They are a much more precise device, and can be rotated more than 20 times faster than a floppy disk. The hard disks are capable of holding a lot more information than a floppy disk. Standard sizes are 10 megabytes, 20 megabytes, 50 megabytes, 70 megabytes, and 100 megabytes. It is not at all unusual to find even small microcomputers with disk drives that can store a larger amount of information. Since the hard disk rotates much faster, the access times are much shorter. Access times are on the order of 10 times as fast. By putting multiple heads around the periphery of the track, you avoid having to wait for a full rotation to get to the next piece of information. For example, with two sets of heads 1/2 a rotation would do.

## Virtual Memory

A the name implies, virtual memory is memory that appears to be there but isn't. Typically, what happens is the following: the RAM memory, which is the only part of the memory to which the arithmetic and logic unit (ALU) has access, is broken up into pieces of some sort - sometimes they are called segments, sometimes they are called pages. Suppose one had a program that was so big it wouldn't fit in the RAM memory, one possible solution would be to take the program and break it up into pieces that are the same size as the little pieces that the memory is broken up into. Then if we had a big secondary memory, a hard disk for instance, we would break our program into pages and put pages 1-6 in one memory and pages 7-12 in another. Now, the program obviously is not all in the fast memory at the same time. The arithmetic and control units can start accessing instructions and there is no problem until the program gets to that end of that segment or page. All of a sudden it is signaled that the next instruction is not available. It is in another memory. What typically takes place is that there is a page swap. Page 1 might be replaced by page 7, so we move page 1 out and move page 7 into that space. Now the arithmetic unit can start picking up the instructions that were just read in and continue executing the program. As far as the program is concerned, you can think of the memory as being just one great big piece of RAM. As far as the ALU is concerned, it can also view the memory as if it were bigger than it really is. There usually is some kind of special hardware that detects a fault when something is being referenced that is not present. This switching back and forth is automatically handled by some hardware elements. Those operations are called paging operations. You have to be very careful so that you do not end up thrashing, i.e., jumping back and forth between pages in such a way that most of your time is spent just moving things in and out of the memory, as opposed to doing computing. We have pretty good memory management schemes now, so it is possible to run programs that are bigger than our memory can handle by using special schemes that have to do with fooling the processor into thinking it has more space than it has.

## Storing a Program in Main Memory

In general, we want to be able to read data in, manipulate it in some way without destroying it, and then display it along with the results of the manipulations. That is essentially what a computing machine is about. The part of the machine that does that manipulation is the arithmetic unit, part of the central processor or the microcomputer microprocessor. Eventually, we put the whole machine together with data coming into the input devices, information being passed back and forth between the arithmetic unit and the main storage and the results of data then being displayed on output devices. Sitting off sort of by itself, but with electrical contact everywhere, we have the control unit, the so called CPU. The control unit runs the machine. The control unit is the part of the machine that has to tell the input device, the keyboard for instance, when to accept something. It has to tell the circuitry associated with the memory where to put information that comes in from the outside - which one of those bytes, which one of the addresses is to be used, and so forth. It has to tell the circuitry associated with the memory which memory location to retrieve information from to take it to a printer or a screen to display it. It would have to send a signal to the monitor saying clear the screen off so new information can be input. The control unit has to decide when and what will be done. If the machine is going to handle our particular problem, we have to get our algorithm into the hands of the control unit.

How do we get our program in there? It is clear that if the CPU is so fast, there is no way that we can give instructions by hitting keys one at a time, telling the machine what to do. If we are manually able to put information into the keyboard maybe twice a second, the machine will be doing nothing during 99,999% of the time it has available to do something. The question is, where can we put the program?

In the late 1940s, John VonNoyman of Princeton, and three of his graduate students, one of whom is Professor Burkes here at the University of Michigan, came up with the concept of the stored program computer. All machines that we now are building or have built since about 1953 are of this type. We know that in the end all the information that's in the memory is just a bunch of bits, just a bit pattern - which, if we look a little further, is just a set of capacitors that are charged or uncharged. Somehow we have to represent our algorithm in the form of strings of ones and zeros. In machine language that form of the algorithm is called a program. So what we do is we take the memory of our computer and split it into several pieces, some of which are for storing the data and some of which are for storing the results when they have been computed in the arithmetic unit.

Another part of the memory is given over to the encoded individual instructions that the control unit is expected to execute. Each instruction in machine language consists of an operation code and an address code.

Each different model of computer typically has a different machine language. It is not possible, usually, to take a program in machine language from one computer and run it on another computer.    It is a completely different processor, a completely different set of instructions, a completely different number of locations in its memory and so forth.    That's a disadvantage of programming in a machine's language: as soon as you buy a new computer you have to rewrite the programs from scratch.

## Keyboards

Most computer keyboards, especially microcomputer keyboards, have something that looks like a typewriter in the center.    Off to the right, there usually are some buttons that have something to do with cursor control.    A cursor is just a little rectangle that is displayed on the screen so that you know where you are and if you type a character you will know where it is going to appear.    The cursor control keys are used to move the cursor around on the screen, to pick out certain things to be changed or to locate something on the screen and so forth.  There are also special keys that allow you to insert characters between characters that are already on the screen.    Deletions are usually done with the delete key or backspace key On the IBM PC, there is a set of function keys.    These are normally programmable, which means that if you write a special program, you have access not only to accepting information for letters and digits and so forth, your program can also accept keystrokes that involve the special function keys.    If you choose, you can make any one of these keys represent anything that you want.    If you are writing a letter editor, you might decide that F1 means erase the line, and F2 means erase the next character, and F3 means erase the next word.    In other words, you can decide when you write the program what these codes mean.    This does not mean you can just arbitrarily    put a program in and start playing around with the function keys and know what's going to happen.  You have to have a manual or some sort of "key" to the keys to let you know what each means.    Typically, there are also special control keys which are involved with special commands to programs.    One of them that is very often used, probably the most used key on the keyboard in fact, is called the control key. Quite often one holds down the control key and then hits one of the other keys.    That usually means something special to either a program or perhaps the operating system. These keys are usually used just like shift keys:    you hold them down and you push other keys.

## CRTs or Screens

Display units (screens) operate in two different modes.    One is called text mode, which is normally very fast and is oriented toward text.    Most machines format their screen for 25 lines and 80 columns of characters.    On the IBM PC you can change to 40 columns if you want. There is also what is called a graphics mode.    Grahics mode is quite different.    It is usually very expensive.    It is somewhat slower in many cases, but you get much better resolution, and you must have it in order to draw pictures. Typically, the screen is formed (the image you see) by lighted dots called pixels - picture elements.    Pixels are little dots on the screen that are either lit or not. For color screens, they might be in different colors. The resolution of the screen and whether or not it is color has a lot to do with how much a monitor costs.    The reason is that for every additional one of these little pixels, there has to be some kind of memory assigned. There has to be a bit pattern that says whether it is on or off, how intense it is, what color it is, and so forth. For large, high-density pixel screens, you begin to pay higher and higher prices for screen memory. A high-resolution screen would normally have something like 1024 x 1024 pixels. Notice again a power of two; that means you have one million of these little dots on the screen. That's pretty good resolution. In other words, letters would look real sharp.

On some microcomputers, everything is in graphics mode. Even text is displayed in the graphics mode, and that is one of the reasons why with the MacWrite software or some of the other programs on the MacIntosh, you can display something in Old German twice as big as all the other letters very easily. This is because this display is essentially drawn. They are not simply characters which have drivers that display them individually and of uniform size. In the future, we are probably going to see more and more shift toward graphics mode for everything, such as we see on the MacIntosh.

Currently, the machines that are at the low end of the market, and especially the IBM clones and the IBM equipment, operate two quite different modes. You have to physically shift back and forth in many cases between the two modes while operating. When writing a program, for instance, you must actually give instructions to switch the mode of the display from one form to another.

## Printers

Dot matrix printers are the typical printers that are used in small machines. We can print either 80 or 132 columns on the standard printers. The IBM PC can handle up to three printers at the same time; that is, you can have a printer 1, printer 2 and printer 3. You can attach all sorts of extra boards in the machine. The IBM PC has what is called open architecture, which means that IBM has published all the specifications for how signals are processed by the machine so that independent vendors can design their own boards which can then fit into the overall scheme of things. This means you have the privilege of buying non-IBM equipment and installing it knowing that you will not damage the equipment and knowing that it will function (if it's a reliable manufacturer). The boards typically fit into some open slot where there are sockets that have been manufactured by other companies or IBM itself ready to accept them.

## Ports

Usually, the computer has some connection to the outside world. The standard connection to the outside world is called an RS-232 port. It is a fairly slow data rate board. The RS-232 port is called a serial port. This means that the bits stream out one after another as opposed to going out in parallel

Many machines have parallel ports on them as well. Lotus printers are driven by parallel ports. This means, for instance, that there might be 8 lines going out simultaneously, representing one byte. A whole character would go out at the same time as opposed to having the 8 bits string along one after another from a serial port. This way the information flows more rapidly.

## COMMUNICATIONS

Next we will consider communication to the outside. The RS-232 port is a fairly slow port. Typically, the fastest rate that you can get for sending those bits out in serial sequence is about 19,600 bits per second. There is a name for that - it's called baud. It is not exactly bits per second, but it is very close to that. To determine how many characters are being transmitted, you can take the baud rate and divide it by 10. If you have the type of device that allows you to use the telephone network for transmitting information (a modem) you will find there are standard baud rates. With a 2400 baud modem, you can ship over the telephone lines about 240 characters per second. That is pretty fast. It  is not so fast that it fills the screen instantaneously, but it is fast enough that you can see the screen filling fairly rapidly. Over standard voice-grade telephone lines where you go through central switching operations (AT&T and so forth) 2400 baud is about as fast as you are going to get. With special conditioned lines you can go much faster

Most of the commercial networks (such as TELENET) do what is called packet switching. This means the data rates are quite high - much higher than 19,000 baud. The information that is sent over these packet switching networks is multiplexed so lots of people can use the network at the same time. The network computers squeeze together little packages of these bits or bytes and then ship them out in bursts. There are lots of people connected at the same time, so associated with each packet there is an address of where it is going, where it came from, some error checking codes, etc.

## NETWORKING

Quite different technology which is becoming more and more important is local area network technology. This means that within some small area, maybe within the range of 1000 feet, we connect together several machines. By keeping them close together we do not have to amplify the signals and have special circuits and telephone lines. A typical data rate for a network is $10 \times 10^6$ bits per second. That is much faster than the 19,000 bits per second that we can send over telephone lines. At these data rates, it is possible to send a picture, the 1024 x 1024 information associated with a picture, fairly quickly. Ordinarily, it would take a long time before you could transmit a whole picture, with all the information about the color of the pixel and whether it was on or off and so forth. To get any kind of graphical information moved around, you need very high data rates by comparison with text information movement. The local area network is one way to do this.

One of the purposes of the local area network is to make it possible to share information, databases, financial information and documents without having multiple copies. One of the bad things about having individual computers is that everybody gets a diskette with the latest specifications on it. If somebody changes a specification, perhaps only one or two people find out about it. Everybody else working on a problem is still working on the old assumptions. On the other hand, if you have a hard disk and you only have one copy of the specification, the changes are made there and as people access the information they get the latest specifications. This minimizes that amount of misinformation. Also in many cases, you have better control of the information. You may not want too many people to access some information, so you put it on the hard disk and then have a password, allowing only certain people to know that password. Salary data may need to be available to certain people, managers and so forth, but not to the rest of the clerks who are also using the system for doing document preparation.

When using a network not only can you use the computing capability of your own local node - what is sitting on your desk - but you can send work to a node that does nothing but computing. That node will work the problem while you do something else at your node. When the computing is done, you pick it up. So not only do you share files and archive information and printers, but you can even share computing services on the network.

## OPERATING SYSTEMS

Let me tell you just a little bit about how we communicate with the computer - what makes the computer usable for the user. What makes it usable is that someone has written a set of programs which eliminate a lot of the difficulty in communicating, for example, with disk drives. Those programs are called operating systems. They are simply programs somebody wrote. The program cannot be used unless it is loaded into the machine's memory. One of the first things you will see when you buy a computer is a DOS diskette. DOS stands for Disk Operating System. It is an operating system that handles disk file drives easily. You will get a few instructions that say, put the DOS disk into the left drive (A drive), and push some buttons or turn on the machine. If you hear some beeping, what is happening is that some programs are being read from that disk drive into the machine's RAM memory. The programs that are being read in are the supervisory programs, the DOS command programs. Now give DOS a set of commands. For instance, one of the commands might be CLS command, that is, the clear screen command. Doing this inputs that command to a program that is now resident in the machine's memory. Otherwise it could not be processed. Those characters, C followed by L followed by S, could not be interpreted if there were not a program in the machine's memory. That is called a DOS command. The user of the computer had to learn the language that whoever put together the operating system came up with. Every machine, unfortunately, tends to have its own operating system programs and its own command characters. So when you get a new computer, one of the first things you have to learn is the command characters. Fortunately, a lot of makers have taken up the so-called IBM DOS, or sometimes it is called MS DOS. (Micro Soft DOS) because that company actually wrote it, and it is used on a lot of machines. All the IBM look-alikes use DOS, so it is pretty easy to move from machine to machine in that class.

If you go to the MacIntosh, it is a completely different kind of operating system structure - in fact, a completely different way of communicating with the operating system. It does not involve typing in commands. It involves pointing to things on the screen, using a device called a mouse, which is just another input device. The heart of any operating system is the manipulation of information. The information is almost always kept in files. Files are simply strings of characters or, actually, strings of bytes or strings of bits. What you put in those files depends on your purpose, but one thing you might want to put into the file would be text. If you are typing a letter and you want to save it on your diskette, that uses a particular set of bytes. In addition to text files, there are program files. Program files are files that contain programs already in the machine's own language. In order to run those programs, all we have to do is read them from the diskette into the memory of the machine and then tell the control unit to go there and start interpreting the instructions. The DOS programs themselves are written in machine language. That is why as soon as those programs were loaded they were ready to accept commands like CLS. There are also files in DOS called batch files. These are text files which contain commands. So if I put a CLS in a batch file, rather than type it on the keyboard, I could tell DOS, to go look in that file and execute all the commands that it sees in that file. That is called a batch file. These are basically the three kinds of files.

The files are named in a simple way. Files have up to 8 characters in their names, usually letters and digits, followed by a period and an extension, which can be up to 3 characters. you might have a problem and you call it PROBLEM 1. Maybe you have some text (maybe it is a statement of the problem) and you call the file PROB1.TXT, just remind yourself that it is a text file. If it is a Fortran program, then you might say PROB2.FOR, to remind you that it was a Fortran program. By and large, you can name files anything you want. Typically, the program files have special extensions, EXE, for example, meaning it is an executable file versus a standard one. So you will quite often see EXE as the name of the file. There is a directory command, for instance. If you give the command DIR you would get the names of all the files that are present on the disk that is in drive A, for example. There is a sort command, such as SORT.EXE. I can say something like this: I would like to have the directory, but then I would like to sort it. If I don't say anything else, it will sort it starting in column 1. Eventually it should give me the directory in sorted order. The A's are followed by the B's followed by the C's and so forth.

There are certain commands called resident commands which can be executed almost instantaneously because they are actually incorporated into the program that was loaded at the very beginning. They are really an inherent part of the DOS command files. the operating system that was loaded. A DIR was one of those commands. SORT is not. Where did sort come from? SORT is in one of those files. A program actually does the sorting. In order to do the sorting, we end up with a set of transient commands. A transient command is a command whose program actually has to be read from outside into the memory before it can be executed. That is one reason it takes longer to sort than to do a directory. The sort program had to be physically read in first before that activity of sorting could be executed and then the output displayed. Just about anything you might do with files, you can do using the operating system.

Machine language is the bit pattern language. It is the only language the computer actually understands. Eventually, everything a computer does must be in the form of the machine's own language. Otherwise, nothing can be done with it. You cannot just instruct the computer to do this or that through the Fortran program. It won't understand it.

## PROGRAMMING LANGUAGES

After the development of machine language came the development of assembly languages. Assembly languages are almost like machine languages except that symbols are used instead of bit patterns. Assembly language would allow us to do things like load in register 1 the value of X, where X refers to some memory location. Then add to register 1 the content of Y. That action might restore from register 1 into a memory location Z. The computer does not understand things like L followed by O followed by A followed by D. It only understands 10100101. So somehow this program has to be transformed into binary form before it can be used by the computer. That form is called translation, meaning that there must be another program already written in the machine's language which reads a program as data and produces a program in machine language as its results. For an assembly language program, the program that does that translation is called an assembly. We don't have to move very far away from assemblers until we get to much nicer languages, like Basic and Fortran, which allow us to put in statements like WRITE X Y AND Z or SET Z = X + Y and so forth. It is not a very long step from primitive symbolic languages to procedure-oriented languages. In procedure-oriented languages, we normally think in terms or in graphical form and then we transcribe it into a sequence of statements which follows some language rules. The language rules are practical rules that involve symbols or addresses like X, Y and Z and some English words like write. Some special operators like = and + and / and separators like ( ) allow us to describe our output without reference to what is really going on at the hardware level. The assembly language is more closely tied to the machine language, because for every machine instruction, there is a corresponding symbolic instruction. With procedure-oriented languages, we get away from that completely. The program that does this translation for procedure-oriented languages is called either an interpreter or a compiler. You may have heard about Fortran compilers. That means somebody has written a program in Fortran and then had it compiled or translated into machine language and then executed it. If you write a program in any language except the machine's own language, there is a two-stage process: first, creating the symbolic version of the algorithm and then having it translated, and second, executing the translated version of the progrm by the processor.

## COMMAND LANGUAGES

Another level of language is the command language. By and large, command language is a language which is interpreted by the operating system programs. Another class of languages is a very high-level class of languages which are usually called problem oriented. The program is really a description of the problem, as opposed to a procedure for how to solve it. For example, you might have a piping network analysis program. The input to the program would be a description of the network of pipes and pumps and elbows and all the other things you have, as opposed to the details of how to calculate a pressure drop between this node and that node. That would already be incorporated in the form of procedure-oriented programs that have been translated, but as a user of the program you would describe the process as opposed to the algorithm for its solution.

There is really a whole hierarchy of languages that you have to deal with if you use computers. Virtually no one deals with the computer at the machine's own language level anymore. but everybody, in order to use a computer, has to learn at least the rudiments of the commands of the operating system command language. Even if you use purchased software, you have to use the command language. The software manufacturer will tell you, "Put the disk in drive A, type the name of the file," or something like that, and then, "push certain keys." That is all part of the command language for the operating system.

## OTHER COMPUTER TOOLS

What other kinds of languages do we have to deal with?  What kind of software do we need?  Let us categorize things:

1) Office Tools-- things that every office has to have.
2) Engineering Tools--things that you need if you are going to do engineering calculations and keep track of things scientifically.
3) Laboratory Tools--if you want to log data in your plant or control part of your plant.

## Office Tools

What kinds of office tools do we need?  We need editors and word processors.  These essentially are programs written in the machine's own language which allow us to input text and manipulate it.  I'm sure you have all seen people in your office using word processors.  It typically takes two or three hours to learn how to use one.  You can become proficient very quickly, and anybody who doesn't use one and has access to a machine is wasting a tremendous amount of time.

It is very easy to edit, you can go back and move things around very nicely.  Spelling verifiers are available.  If you are not a very good speller, you can take your processed text, pass it through a spelling program and it will point out all those words that it thinks are misspelled.

Database  managers essentially allow you to take information and format it in the way you would like to format it.  It is usually in record form, where each record has a set of fields that are associated with it.  A record might be a personnel record and the fields might be name, department, salary, number of dependents and all the other things that would be associated with a record for a person.  Then the data manager program would allow you to put information into a database which can be modified as necessary.  It is also easy to interrogate the database.  For example, say "I would like to find all the months when the water bill was greater than $25.00." It is very easy to use conditional records and pieces of records to obtain selected information.  The system can very easily narrow things down, prepare mailing lists, keep information up to date, and create new databases from old ones.

Spreadsheets are simply rectangular arrays of numbers, text, and in some cases pictures that you can manipulate in very regular ways.  It is amazing what you can do with spreadsheets - incredible things.  Spreadsheets are probably the major use as far as microcomputers are concerned.

There are essentially three different kinds of graphics.  First, there are some business bar charting kinds of graphics usually called presentation graphics, pie charts and things like that.  They work pretty well even on fairly low-resolution screens.  Second, there are some technical graphics that involve plotting.  Those take better resolution.  In other words, you don't want to plot with all kinds of jagged characters and so forth in it.  Third, there are interactive graphics which allow you to draw something.

## Engineering Tools

For general engineering work, you need numerical methods and a statistical analysis program - the sort of things that are used to solve technical problems. To solve a system of equations, you need engineering tools like this.

Suppose, for example, that one of the elevated tanks is emptying. What happens? Suppose a pump fails. What will happen in the system? Perhaps you want to plan some additions to the network. What will this do to pressures elsewhere in the network given certain demands? What locations in the network will be affected? These are by and large those kinds of programs that are problem oriented. You describe the network, as opposed to putting in the Manning equation for getting pressure drops and all that sort of thing. Chemical process simulators and design programs are other engineering tools. There are just an enormous number of these things available. Some of them are in the public domain; most of them are commercially produced, and you have to pay for them. The better they are, the more they cost.

## Laboratory Tools

Finally, there are a set of laboratory tools which are usually combinations of hardware and software. These have converters that will, for instance, take analog signals from thermocouples and convert them to digital form so that they can be processed inside the machine. The machine only deals with digital numbers. We have the capability of both accepting signals and sending out signals. Part of what we want to do is data acquisition. We want to monitor what is going on. The data acquisition consists partly of logging the data. This might involve keeping the data locally on the data acquisition board. There might be memory. There will almost always be its own processor on one of these boards. It might mean keeping things in the memory of the machine, the main memory RAM, or it might mean writing it out on one of the disks. There may be simultaneous analysis if the machine is fast enough. As the data come in, the analysis is done, the regression lines are fitted, and so forth. On the other hand, the data may simply be logged, put away and later on read back in to the processor. So there is a combination of software and hardware that almost always goes together when you are using laboratory tools. At the sort of upper end of the laboratory picture, we have the taking of output from the computer, converting it to analog form, or in some cases sending out digital signals directly to equipment, which will then change all settings - all the things that are related to the operation of the plant itself.

## SUMMARY

A computer is actually a system of connected pieces of hardware which can take in data, process it, and produce usable results. In order to do this the computer uses specialized software called operating systems which handle the communication between the components and the control of the processing.

Application programs are written to do specialized processing for an area like engineering. Networking and communication software allow many computers to be linked together into one larger system.

As computers become more ingrained in our society, we must all have acquired a basic understanding of computer concepts. From there we can go on to study any more specialized area of computing.

# CHAPTER 2

## SELECTING A COMPUTER AND SOFTWARE:
## A USER'S VIEWPOINT

Dan Phillip Wolz
Wyoming Wastewater Treatment Plant
Grandville, Michigan

## INTRODUCTION

Please join me as I relate experiences I have had learning about and trying to purchase a computer. The first section is titled "A Learning Adventure in the Computer World." The second section is titled "Making a Family Decision in Wyoming, Michigan." A few conclusions will be drawn from these events.

Most of what is presented here was learned as my co-workers and I struggled to make some sense out of a whole new field to us. I would recommend preparation as a major key in the purchase of a system. It is overwhelming at times. It is not impossible.

Imagine, if you will, if the tables were turned and a computer programmer were faced with terminology such as: mean cell retention time, trihalomethanes, chlorine demand, cyanide amenable to chlorine, gram negative, nonspore forming, and dozens more. It then may begin to be seen why I have responded as I have in this chapter to the new terminology confronted in the field of computerization. Please note the term "boss" is used generically and affectionately to mean any one of the management team.

## A LEARNING ADVENTURE IN
## THE COMPUTER WORLD

The adventure started with some preparation and various questions which may be of some help to anyone wanting to purchase a system.

### Preparation

Before a system can be configured at a facility, preparation is necessary. It does not matter how many minicourses are taken or how many books or magazines are read, these basic questions must be answered in some manner before hardware and

**19**

software can be selected.

1.  What do I do?  Determine if your duties can be computerized by looking at what you do.  A computer and software can reduce the time it takes for many of the more routine parts of your job such as:  calculation, budgeting, inventories and payroll.

2.  What do I want a computer to do?  A computer properly programmed can do just about anything; but, if your application is exotic or unique, you are going to pay for it by special software and people to maintain it.  Do not fall into the trap of going to your local computer store and asking, "Can it be done?"  The answer is always, "Yes."

3.  What are the costs?  After deciding what to do, look realistically at what the market is offering and see if it meets your needs. This includes reputable firms and consultants.  Be ready with several alternatives.

4.  What are the benefits?  Look at the possibilities and variety of things that a computer system can do not only for you, but for your boss.  If the boss is the least bit interested, look for the things that will benefit him.

5.  How can I get the boss to see the benefits despite the costs and not say, "No"?  First, deal with all of the alternatives and be prepared to be flexible.  Think larger than your actual needs so that if something less than the desired system is discussed, you will still have some choices.  Second, present the items that the boss is thought to be interested in.  Do not tell him about linear regressions if he is interested in whether or not it will print commas and vice versa.  Third, tell the boss that there are other things that can be done with the time saved.  Do not tell him it will save labor costs.  It may, but more realistically it may allow you to delay hiring additional personnel, as opposed to laying off existing employees.  Finally, if you have convinced the boss, you may have sold the city.  At least you have gone a long way in getting your system.  Make sure all your correspondence to the boss is written so that his boss or bosses can understand it even though it may be sent only to your boss.

## Hardware

Next, after the major questions are answered, look into some of the hardware requirements.  What is hardware?  Hardware is everything, including the floppy disk, which allows the commands recorded on the floppy disk to perform a function.  For example, a printer will print, but not until it is directed to by a print command. The printer is hardware, the command is software.

Hardware needs are often emphasized to a greater degree than software needs. This is a mistake.  Decide what task is needed and design systems needs around software that will do it. All hardware does is provide the means of letting the software operate. However, most software is designed to be compatible with most hardware. There are just a few things hardware does, but the language and "buzzwords" which have developed around this field are overwhelming to anyone just beginning to explore it.

After all the verbiage is separated, all that is left for hardware is: Input, process and output. "Input" is all the external commands given to the computer mostly by keyboard or joystick. "Process" is the manipulating part of the three part process. It calculates, sorts, files, arranges and performs a variety of other functions. "Output" refers to the results of the manipulated commands and their display.

Computers come in different sizes. The sizes are divided into three types. They are, from smallest to largest, micro or personal computer, mini, and mainframe. Generally, the larger sizes can handle and manipulate more information faster and conduct several functions at the same time.

## Operating Systems

It takes a certain set of commands to make a computer operate, and every computer must have an operating system that makes it go as well as a language that it understands. If the operating system is specific to the computer brand, then programs must be designed to use the operating system for that computer. Operating systems tell the computer how to electronically code information. Even though it is all stored in binary, it still needs a way to retrieve and use the information. Examples of operating systems are: Disk Operating System (DOS) and Control Program for Microprocessor (CPM).

## Language

The language a computer uses can be changed to adapt to the task to be accomplished. It is the way the operator communicates with the computer. The rules of the language must be exact and usually are used by expert programmers to design special programs. Usually the language is an acronym of a larger phrase. For example: Fortran stands for formula translator and is used in scientific application. This language is one of many. Fortunately, for the initiate or "nonuser unfriendly" it is not necessary to know much about these languages unless he or she is going to design his or her own software.

## Software

Software is specific programming to run the tasks which were decided by answering the "big" question which was already discussed. The question again is. "What do you want your computer to do?" The answer will determine specifications for your system.

There are also two questions which will help determine what type of system software will be needed.

1.  Is process control required? Traditionally process control has been very expensive because of mainframe equipment requirements, but it is very powerful and can support multiple users and simultaneous tasks. Usually a programmer is required to operate the system. However, consultants are sometimes retained to develop and maintain software. There are many new products coming on the market that require only personal computers (PC's). These can do many things, including partial process control, for a much lower cost. There are merging technologies which blend hardware and software applications to wastewater.

2.  Is data processing the main requirement and are calculations along with word processing to be performed at the same time by different users? This may require a mini system to support multitasking and simultaneous use. However, software for mini computers is not as cheap as for a micro.

If specifications are required at a facility, make these performance oriented, especially if the plan is to buy "canned" software. Make the vendor demonstrate his programs and write requirements into the specification. Include training in any specification. Good specifications should also include the cost of a maintenance or service agreement.

## MAKING A FAMILY DECISION IN WYOMING, MICHIGAN

In this information, I would like to refer to elected city officials as "city fathers" or better yet "Dad." And the taxpayers I would like to call "Mom." You see, with a parental structure like this it puts any who work for treatment plants supported by taxpayers and governed by elected city officials in the role of children. All are part of a family and are really "brothers" and "sisters."

Be reminded again that when the term "boss" is used, it is used generically and affectionately as though all of my bosses were older "brothers" and "sisters."

And as in most "families" things may look quite normal, but may be quite different behind closed doors. So as I describe my "family" circumstances, a little underwear may be publicly exposed, which normally would not be presented to other "families." Keep in mind the intent is not to hurt my "family," but to demonstrate how this "family's" decision was made. I will try to be as generic as possible, not only because I have the highest regard for my "family," but because I value my job as well.

With the exception of the historical part of this decision, this experience is mine and from my point of view, as from a "son" to a "father." I therefore could not know all of the reasons why my "Dad" has made his decisions and the timing in which they were done, but "Dad" is older and wiser and I abide by his decisions.

The story of the city of Wyoming and computers can trace its roots back as far as 1969 when initial contacts were made by various vendors, and some of my older "brothers" and "sisters" began to catch a glimpse of some of the fantastic possibilities for computers to handle data. A few years later, at the recommendation of our "family's" accountant, a computer consultant was hired to recommend a system and software for the Water and Sewer Billing Department, Accounting Department, Purchasing Department and eventually other departments. "Dad" liked this idea and approved the recommendations for our "family" to purchase a computer in 1979, the same year I was born to this "family."

Water Billing was the first department on the system.    Any problems with software were handled through the consultant.    No programmers were hired.    The philosophy was to "customize" and "patch" software, as needed, through the consultant.    Existing city staff was trained in operation commands of the software but not in designing programs.

About the same time, much interest was sweeping the country about the use of personal computers, or "P.C.s."    Initially, as everyone knows, the main use of these P.C.s was for playing electronic games.    With the stigma of a "toy for games" attached to a P.C., "Dad" was reluctant to even allow any of his "children" to purchase P.C.s.    There are two notable exceptions to this.    The first was that a P.C. was purchased for use in the clerk's office for mailings and other things, but was not used much and was in storage for a while.    Finally, when it was taken by the Motor Pool Department for use, many games were found on it, thus proving "Dad's" reluctance to purchase.    It is now being put to good use, however.

The second exception to the "anti P.C." feeling by "Dad" was the purchase in 1983 of a two terminal P.C. system for the Water Treatment Plant.    However, after it was approved by "Dad" it was discovered that two terminals could not run from programmed floppy diskettes and one terminal had to remain idle for almost two and one half years.    The Water Plant "brothers" and "sisters" made tremendous accomplishments despite this problem.    Maybe "Dad" let the Water Plant have this P.C. because he liked them "best."    In reality, it was probably an experiment to see if maybe the other departments ought to do the same thing.

Meanwhile, back at city hall, the second department was put "on line" to the mainframe computer.    All went fairly well.    This addition slowed the system down. About the only thing that most of the "family" noticed, however, is that before they went "on line" payday was on Thursday and after going "on line" payday was one day later, on Friday.    Most of this "family" history was unknown to me and my "brothers" at the Wastewater Plant before we readied ourselves to approach "Dad" with our requests.    This history was relayed to me by my older "siblings" to the best of their recollection.

One thing we did know was that we wanted to look good to "Dad" and look as if we knew what we were doing.    Naturally, we asked for more money ($14,900) than our "brothers" at the Water Plant.    A little "sibling" rivalry.    In order to present an educated choice, two older "brothers" and I enrolled in a computer literacy course to try to gain some understanding of computer terminology.

At first it seemed difficult, almost as hard as trying to talk to a water operator about biochemical oxygen demand or a wastewater operator about turbidity. But by the sixth session we were able to design and operate our own small programs. I am sure the major software designers were not worried in the least about this new-found knowledge.    After graduation from the class we felt confident enough to engage in conversation with any over-the-counter sales representative in the area.    Over the next month (April 1983) at least half a dozen multibrand dealers were visited.    At the end of that time, we were sure we could get just about anything we wanted for our $14,900.    In May 1983 we had found one dealer who would provide us with a four terminal multiuser system with two printers and furniture.

During the months of May and June of 1983, after continued preparation and planning, we felt we were ready to purchase.    The money was approved and the specifications were prepared in readiness for July 1, 1983, the beginning of our fiscal year.

Next, the scene shifts as the city's administrative assistant, an older "brother," is asked to work with us. After informing him of our progress to that point, he said we ought to spec software and not hardware. I said, "What about furniture?" He said, "Do not worry about it. Write a spec for performance not for hardware." I said, "Okay," and was handed several performance specifications from other cities to use as examples. My original specifications were discarded and I started over.

By January 1984, the new set of specifications was complete and just in time for another budget request. I requested another $10,000 for equipment to add to the system we had not bought yet. Over the next few months "family" discussions continued and the specifications were sent to city hall for preparation to bid. These specifications were performance oriented and mentioned very little about hardware and nothing about furniture, but all the tough questions were answered. We were prepared. Our bosses were prepared. The money was there, ready to spend. We were all set except for the actual bid process, which was next.

However, the month of April 1984 was one of the major disappointing times in this effort to buy a computer. First, the extra $10,000 requested in January was turned down as a budget tightening move and because of the flux in deciding what to do about computerization city wide. Second, "Dad" flatly rejected requests from the Water Plant for equipment to make their second terminal operate and from the Purchasing Department to buy a word processor. These requests were called "toys" by "Dad." "Dad" wanted a study done. His main question was, "Why are these departments not connecting to city hall's existing mainframe as called for in the original plan?" "Dad's" question was fair. As a result, a moratorium was placed on any and all future computer equipment purchases until a study was complete. We never even got a chance to get the specifications out to the vendors. The existing consulting firm was chosen to do the study since they told "Dad" they would do it for free.

One month later, to complicate matters further, the administrative assistant quit to take a new job. The next six months were spent stunned, waiting for someone else to take his place and wondering if the replacement would have the same attitude about performance specifications.

The study was completed in December of 1984 and listed the requirements for each department. Assuming that I would need it "someday," I followed up on the study and priced system needs for the Wastewater Plant.

One month later, January 1985, it is budget time again. Using the prices I had just obtained, $20,000 is requested for hardware and $17,000 for software, just in case the previously approved $14,900 is not carried over for the second budget year.

It is also an election year and about the only time I ever see "Dad" get nervous. Surprisingly enough, these new study-based budget requests were approved in June 1985 by "Dad," and the $14,900 was carried over from the previous year. Nothing further was done about our computer system until October of 1985. But what was happening was that "Dad" was involved in a lot of political and election campaign activity.

Over the summer of 1985, as campaign activity began, a major controversy developed over the residency status of one of "Dad's" challengers. Public focus was shifted to this election. So was "Dad's." Routine city business was left to sideline attention.

As part of the "fallout" of this "family" squabble, some of my older "brothers" and "sisters" determined that it might be a good time to try to buy computers, since "Dad" was preoccupied with the election. There was a good chance he might approve just to avoid any further squabbling and this would not look good just before the election.

On October 1, 1985, I was directed to bid the hardware and software that I wanted and to bid exactly what I wanted and not use such famous phrases as "equal to or better than." I spent the next two days updating prices and preparing another set of specifications. During preparation I said, "What about furniture?" I was told, "Do not worry about it, no problem, but do not include it now." In order for this to get done before the election, I had to get final administrative approval and get it through the Purchasing Department to get it to "Dad's" business meeting before October 28. It seemed impossible. If "Dad" said okay, it could be put on the agenda for approval on November 4, 1985 (election eve), the next business meeting.

I went to the new administrative assistant two days later and told him what I had been directed to do and the timetable. His orders were not the same as mine and it took further direction from an older "brother" to decide what to do. The decision was to take all of the utility computer needs and bid the equipment before the election.

Now that administrative approval had been won, to make it legal according to our city code before October 28, I had to get it through the Purchasing Department by Friday, October 6. It was Thursday, October 5.

Here is what I heard upon entering the Purchasing Department: "You cannot get something out for bid in less than 10 days. How can you write a closed set of specifications? It is like saying you want to bid a car, but it has to be a Chevrolet." My responses were: "Do you want to place the 'invitation to bid' ad in the paper tomorrow or should I do it?," and, "I am saying I want to bid a Chevrolet, but I want all the Chevy dealers to bid on it."

I had a complete set of specifications prepared, but they wanted to change everything. They did not. I even had an ad written for the paper. The Purchasing Department finished it by Friday the 6th, with all bids due by October 22, 1985. And, also bid at the same time, was a duplicate system for city hall's computer.

After the bids were opened on October 22, 1985, selected, and recommendations prepared for "Dad's" first discussion meeting for October 28, 1985, it was decided to have several "brothers" present (to talk to "Dad") at the business meeting along with one of the bidders to answer any technical questions that "Dad" may have. There were no questions and "Dad" said it would be scheduled for the final meeting on November 4, 1985 (election eve).

November 4, 1985 came, and not only was our computer equipment approved, but the remaining equipment for the Water Plant and the water billing computer for city hall. It was all approved as one agenda item. Not one question was asked, although several "bosses" were present along with the computer representative. "Dad" had come through. Thanks, "Dad."

The next day I ordered the equipment, but there was no furniture on the order. I brought this up again and said, "What are we going to set this system on?" The answer, "Do not worry about it. We will get it when the equipment arrives." No problem, I thought.

On December 26, 1985, the system started arriving, and arriving, and arriving, and arriving. I am now the proud owner of a $42,500 system consisting of five computers (one mini, four P.C.s), $10,000 worth of software, three printers (two letter quality, one dot matrix) and approximately 100 operation manuals piled five feet deep. And what about furniture? No problem. The equipment is all setting on cardboard boxes. but I am not worried about it.  I know "Mom" and "Dad" and the "family" will take care of me.

## CONCLUSION

What have I learned?  I have learned that I am a small part of a functioning "family."  That no matter what I think about "Dad's" decision. I will abide by it.  I must.  Besides. my older "brothers" and "sisters" and "Dad" are trying to do what is best for not only me. but for the whole "family."  Also. if I wait long enough "Dad" will probably give me what I want not necessarily because I want it. but because he may be more preoccupied with other "family" matters or he is just sick of my asking.

Lastly. there are some advantages to waiting.  For one. the company that was originally contacted and quoted us hardware. software and furniture is no longer in business.  In fact. the computer is not even made any longer.

There have been four market cycle changes in the last three years.  This means that every nine months the market is coming out with new products and technology, forcing those companies with low market shares to stay competitive.  With that kind of volatility in the computer market the products get better and the price goes down.  As we. the users. make our choices. things begin to stabilize around the software and hardware to run it which we believe will best suit our applications.

# CHAPTER 3

# USE OF COMPUTERS
# IN WATER SUPPLY REGULATION

James K. Cleland. P.E. and Karen Kalinowski
Division of Water Supply
Michigan Department of Public Health
Lansing, Michigan

## INTRODUCTION

The first question one might ask is, "Why is it necessary or important for the State of Michigan to invest in large, expensive computer systems in the regulation of water supplies?" There are many good reasons, some not obvious to the public or a person working in the water supply industry representing utilities, consulting services or vendors of products used in the industry. We will explain not only why computer systems are valuable in water supply regulation, but also why we employ a combination of mainframe and microcomputer systems to best meet our data handling needs.

The application of computers in state regulation falls into three broad categories: those applications resulting directly from statutory requirements; those which directly assist program activities and provide management information both in and outside the agency; and those which are projected but as yet are not in service.

## COMPUTER APPLICATIONS BASED ON STATUTORY REQUIREMENTS

The Michigan Department of Public health supervises all public water supples under Act 399, P.A. 1976 and is a primary enforcement state under P.L. 93-523, the Federal Safe Drinking Water Act. Thus, even though local water utilities collect and store literally mounds of data relative to their water system, considerable information must be available collectively for all water systems in one format.

The regulatory requirements vary greatly based on the class of system regulated. Community systems are those public water supplies serving over 25 people daily in essentially a residential setting. Approximately 1500 community water supplies exist in Michigan, ranging from the small apartment or mobile home park serving just 25 people to the Detroit Metropolitan Water System serving nearly four million people. Noncommunity public water supplies, numbering more than 11,000 in Michigan, serve nonresidential facilities with public access. such as restaurants, schools, campgrounds and industries. The record keeping requirements and the statutory requirements for community systems are much greater than the noncommunity systems.

## Inventory Data

A comprehensive inventory is required of all systems. The community inventory, of course, requires much more detail. The inventory is provided to EPA and the total number of water systems is used in the formula to determine the magnitude of the state program grant awarded under the federal statute. The inventory is a large file with more than 100 data elements, including name, address, population served, treatment practiced, ownership and water source, to name a few. To illustrate how large these inventory files are, we currently are storing approximately 10 megabytes of inventory information for community and noncommunity water supplies, about one-fifth of our mainframe storage.

## Analytical Data

Monitoring of community water supplies dates back to 1913, when the state first initiated a water supply program. The type of monitoring and frequency of analysis are two critical components of any computer system. Monitoring requirements are apparently dynamic: each new regulation based on the most recent research findings changes the data base needs. The data is in tremendous demand from all types of organizations and the public. The turnaround time from sample collection to reporting presents special demands on utility personnel, laboratory staff, regulatory staff and the computer systems. Violations of standards, either MCL's or monitoring requirements, must be compiled and reported to EPA along with specific follow-up actions by the water suppliers and the state. To give an idea of volume, our laboratory currently runs approximately 46,000 bacteriologic analyses and 67,000 chemical analyses each year from public water supplies, all of which must be tracked to determine compliance with drinking water standards.

## Surveillance and Evaluation

The state regulatory personnel must inspect and evaluate the status of each water system periodically. These evaluations must consider all regulated activities of public water suppliers, such as water quality, monitoring, ability to meet demands, construction and operation standards, operator certification, cross connection control, emergency planning and others. The engineer can track the dates of last visits, the status of noted deficiencies and compliance status through our computer inventory system.

## Operator Certification

Operator certification requirements are explicit in the statute and the state must have all records necessary to determine status of compliance at any time. We also use a data base to store records on examination questions which we use for question evaluation and selection. We currently store records on 2,500 certified water treatment and distribution operators.

## Laboratory Certification

Each laboratory performing analytical work for determining compliance with the statute must be certified. There are presently more than 160 certified labs in Michigan.

## Record Keeping

The federal and state statutes impose record keeping requirements on water suppliers and the state. The records of all the various activities mentioned in this chapter are maintained by the state and most of these records are amenable to computer storage, retrieval and reporting.

## COMPUTER APPLICATIONS DIRECTLY ASSISTING PROGRAM ACTIVITIES

The Michigan Department of Public Health has already established computer systems to assist state personnel in carrying out their work and provide data to other agencies and the public as a service.

## Water Production and Use

Each water supplier providing treatment submits a monthly operation report to the state. This report details all operation records concerning water production, chemical application and in-plant control monitoring. An annual report is required of all other suppliers essentially to detail water production data. Our computer water system inventory will retain and report this information and keep a historical record for staff review.

## Administration

Much of our administrative work load is now performed on microcomputers. Budgeting, records of work activities, personnel lists, program plans and word processing are all maintained on the eight microcomputers located throughout the division.

## Operator Training

A large data base for carrying out operator training functions has proven to be an essential adjunct to the operator certificaiton program. We must have accurate records to verify operators' qualifications to become certified. We use the computer for registration applications in training courses co-sponsored by the department. We are also tracking the training course attendance records of all existing certified operators to determine if recommended continuing education standards are met.

## Special Reporting

The state receives literally hundreds of requests for collective data on public water supplies. Many of these requests relate to water quality data, but the uses of collective inventory and analytical data are virtually unlimited. Our own personnel frequently want special reports for data to support their field activities. Breaking down collective data into sub-groupings for comparisons and evaluation is a common need and a desirable feature of a computer system.

Special report requests originate from other state and federal regulatory agencies; planning agencies at the state, regional and local level; local water utilities; trade and professional organizations; local health departments; consultants; universities; vendors; research foundations; environmental action groups and others. The state has no direct obligation to fill these information needs, but we believe that a great service is being provided by making good use of the data available through computer systems.

# FUTURE APPLICATIONS OF COMPUTERS

As we have painfully grown into computer systems during the onset of high tech computer applications to our society, one problem continues to surface: frequently the operating system becomes obsolete by the time it is operated and used most efficiently. As a user, we are told to identify what we need or want prior to system design. However, users rarely understand the power and capability of available systems and are many times unable to anticipate all the valuable applications of the system. We are thus left with regularly discarding old systems and buying new ones or paying for expensive enhancements if they are possible.

The application of statistical software for analytical data is an excellent example. Although statistical software is widely available and fairly inexpensive, we simply do not have it on line yet in the water program. As a state, we also do not have a system to store and retrieve all water quality data generated by state laboratories, such as the Departments of Public Health, Natural Resources and Agriculture. We also need to relate this quality data with the geological information from well records which are submitted for all new public and private wells.

Health data, disease incidence, data on waste discharges, land use, air quality and other examples, if collected and stored by the state in an accessible data system, would make identification of health and environmental problems much faster and more possible to achieve.

New areas within the water supply field which are not a direct statutory requirement, but if available would provide extremely useful data, are water rates and facility replacement or expansion needs data (infrastructure needs). We hope to begin soon the infrastructure needs evaluation process, complete with data storage, with a goal to update continuously as inspections are carried out by water supply personnel. Hopefully, local water utilities will be encouraged to begin an annual capital outlay program with ongoing future needs assessment and timely adjustments in water rates to pay for needed improvements.

## MAINFRAMES AND MICROS

Some of the computer applications described in this chapter can be satisfied with microcomputers.     All small files, administrative files, program plans, budget, personnel and others are handled very well.     The division has Sperry Personal Computers (IBM compatible).     The primary software packages used are Leading Edge (word processing) and Lotus 1-2-3 (spreadsheet).

The following graph (Figure 1) was created and plotted from a microcomputer using data generated from the large public water supply mainframe computer inventory file. This type of data is in constant demand from outside parties.

Because the public water supply inventory, coupled with water quality data files, water production and use records, surveillance and evaluation information, and facilities data, creates a huge storage problem with a large number of water systems, our experts considered a mainframe application best for this system.

The features of the system considered essential to match the information flow, reporting needs and the character of individual staff users are:

1.   Immediate access for several users simultaneously on large files
2.   On-line update capability
3.   User friendly-menu driven
4.   Flexible reporting capability
5.   Compatible with federal reporting system
6.   High degree of user control on input and output

Our historical experience with a batch system led us to conclude that on-line capability was the only way to properly process analytical results from the bacteriological and chemical laboratories operated by the department.     We also found greater desire on the part of staff to make good use of the system when feedback was immediate.     The fact that technical staff use the system directly also dictated the need for a user friendly system.     The system selected by the department is a Honeywell Level 6 computer, for which we also purchased seven terminal stations and four printers.     These terminal stations are in areas accessible to all division staff as well as the water lab and our Upper Peninsula Office and lab.

Because our reporting needs are so complex and varied, the system provided a framework for handling not only routine, on-going program needs, but also provided a system which we could easily control and modify.     Previously, all reports were requested from a central data processing staff.     This procedure was expensive, was not timely in meeting needs and required considerable paperwork.     Our present system makes use of several standard reports used on a periodic basis from daily to yearly and a software package to format special reports.     We can sort and select any combination of water systems for standard or special reports.     For example, we can readily select only water systems serving over 1000 population in Kalamazoo County employing a ground water source arranged in alphabetic order.     This flexibility has proven itself very beneficial to us.

In the design of the inventory system, the large differences ranging from the smallest systems serving 25 people to communities as large as Detroit gave us many problems.     Within individual data elements, we were always aware of the field size to accommodate Detroit or the total statewide numbers if a field could be summed.     On the other hand, when listing facility sizes like storage tank volume, or capacities such

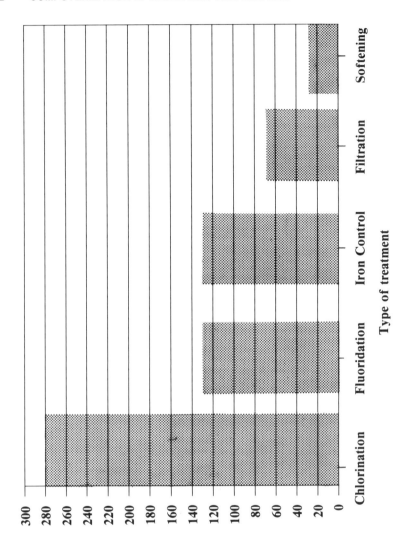

**Figure 1.** Water supplies in Michigan shown by type of treatment.

as pump capacities, we frequently resorted to a code to select an optional unit; e.g., 1 = 1000 gallons, 2 = 1 million gallons. This allowed us to use meaningful numbers on each facility's record.

Another problem was in dedicating space for the various records applicable to each supply, particularly facility description records. For example, records of operating wells might range from none (system not using wells) to more than 150. We found a unique solution.

Each water system has one master record called Part 1, containing inventory items like name, address, population, and ownership. Each master record has a Water Supply Serial Number (WSSN) which makes it unique. In addition, an unlimited number of related records can exist in Part 2 of the system linked only by WSSN. Here are the related records possible in our Part 2 menu:

1.  Ground water
2.  Purchased water
3.  Surface water
4.  Distribution
5.  Treated water storage
6.  Treatment plant
7.  Individual source production
8.  Combined water production
9.  Water use information
10. Remarks line (for narrative descriptions)
11. Monitoring information

The staff member simply identifies the WSSN and another three digit number to place the record in the appropriate location within the Part 2 menu. Remarks can also be placed with any Part 2 record. The final report for any supply will include only applicable Part 2 records, which automatically provides appropriate space for the size of system we are dealing with.

Likewise, when completed, the system will contain records of all analytical data--bacteriological, chemical, and radiochemical--as related records associated to the master record by WSSN. Presently, the chemical records are stored and reported inefficiently in a related data file designed only for small numbers of records.

## CONCLUSION

Computers have many useful applications in water supply regulation and many additional uses will be found in the future as regulations become more numerous and complex. Combining mainframe and microcomputer use can be accomplished; it can be efficient for the varied applications to make use of the best features of each system.

Computer use will only expand in our society as we discover ways to assist our personnel in doing their jobs, in providing and evaluating information and maintaining records. But like many other activities, the value received is in proportion to the time and money spent to design, purchase, and operate the proper system.

# CHAPTER 4

## OPERATIONS CONTROL USING MICROCOMPUTERS

Rolf A. Deininger, Ph.D., M., ASCE
School of Public Health
The University of Michigan
Ann Arbor, Michigan

## INTRODUCTION

Microcomputers are becoming so inexpensive and ubiquitous that no water or wastewater treatment plant will be without this equipment. The range of computers is very wide, from the inexpensive Commodore computers to the top of the line IBM PC AT computer.

There are many uses for microcomputers. Probably the first is word processing. Anyone who has used programs like WORDSTAR, WORD or the IBM WRITING ASSISTANT will never go back to a typewriter. One of the major reasons is the availability of spell checkers which catch the "typos" and also cover the sometimes imperfect knowledge of the English language.

Spreadsheets exemplified by LOTUS or MULTIPLAN are of great utility in organizing storage, retrieval, and editing of the many data originating in a plant, and are ideally suited for summarizing the data in monthly or annual operating reports.

The major emphasis of this chapter is the use of microcomputers to acquire signals from instruments, record the data, and take action based on the signals received.

## INTERFACES USING THE GAME INPUT/OUTPUT PORT

Most of the microcomputers have a game I/O port. This port was the original major device in the sale of microcomputers - the object was to play games with it. The original Apple and Atari computers had (and still have) game ports to which external devices like joysticks and paddles can be attached.

The VIC 20 has a game input port which accepts five buttons (switches) and two analog ports. If a typical VIC 20 (or Atari) joystick is disassembled, one will find on the printed circuit board five bubbles, which when pressed close a circuit. When the joystick is moved forward, a switch closes, and so on. A Commodore 64 has two game I/O ports, and thus allows ten switch inputs. In addition to the switches, there are two analog inputs for each port. An external potentiometer (variable resistor) is examined, and the computer will determine a number which is proportional to the external resistance. Typically, a 0-150K Ohms external resistance is mapped into a number between 0 and 255. Thus, there is a total of $2^8$ (256) possible states in which the external resistor can be. We therefore have an eight bit analog to digital converter.

The APPLE IIe has a game I/O port with three switches and four analog inputs. The APPLE IIc has only two switches and two analog ports. The IBM PC (with an added game I/O card) has two switches and four analog ports. The Atari and RadioShack computers also have game I/O ports.

What can be done with these I/O ports? Basically, any measurement in a treatment plant which can be presented in an on/off (switch) fashion can be connected directly to the microcomputer. Th simplest example is probably a "security" system with two doors and windows having switches with magnets attached. When a window is opened, the switch changes its state from closed to open (or open to closed); this can be directly monitored by the computer. A tipping bucket rain gauge has a switch and magnet which temporarily closes when the bucket tips, and thus gives a signal when the bucket empties after a certain amount of rainfall. A wind direction vane can be represented as many switches. If a relay is put in the power line to a pump, one can confirm the existence of a current flow and presumably the operation of a pump. If infrared emitters and collectors are used, the beam will be broken by water and specific water levels can be determined. There are many more possibilities.

If the analog input is used, several devices are suggested. A thermistor - a resistor which varies its resistance with temperature--is ideally suited to measure the temperature at any point in the plant. A photoresister - a cadmium   sulfide cell - varies its resistance as a function of the light which strikes the cell. Thus it can be used to measure light level, confirm that a light has been turned on, or measure the level of a sludge blanket.

## Programming a Game I/O

In order to use a game I/O port for the above uses, two things are necessary: (1), the "pinouts" of the connectors, and (2), the specific statements in a programming language needed to determine the status.

For example, Figure 1 shows the switches and analog inputs for an APPLE II computer. A simple program in BASIC that would query the status of switch 1 follows:

```
10 S1 = PEEK (-16287)
20 IFS1 > 127 THEN PRINT "CLOSED"
30 IFS1 < 127 THEN PRINT "OPEN"
40 GOTO 10
```

Similarly, a thermistor connected to pin 10 and pin 1 could be read as:

```
10 A = PDL (1)
20 OHMS = A x 150/255
30 PRINT "KOHMS = ", OHMS
40 GOTO 10
```

The internal RC (resistor-capacitor) circuit in an Apple does not always correspond to the exact 150K Ohms and needs to be calibrated, but the readings are repeatable.

Figure 2 shows the pinouts and the necessary BASIC program to read the inputs to a VIC-20 microcomputer.

Switches:

```
S0=PEEK(-16287)
S1=PEEK(-16286)    if answer >127, switch is closed
S2=PEEK(-16285)
```

Analog Input:

```
PDL(0),PDL(1)    external 0-150 k Ω resistance
PDL(2),PDL(3)    converted into reading of 0-255
```

Switch Output (annunciators):

```
A0    POKE -16296,1    off
      POKE -16295,0    on

A1    POKE -16294,1    off
      POKE -16293,0    on

A2    POKE -16292,1    off
      POKE -16291,0    on

A3    POKE -16290,1    off
      Poke -16289,0    on
```

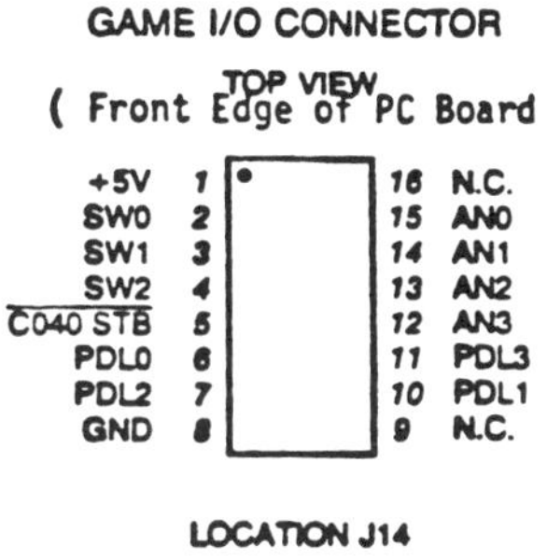

GAME I/O CONNECTOR

( Front Edge of PC Board — TOP VIEW

```
+5V      1        16  N.C.
SW0      2        15  AN0
SW1      3        14  AN1
SW2      4        13  AN2
C040 STB 5        12  AN3
PDL0     6        11  PDL3
PDL2     7        10  PDL1
GND      8         9  N.C.
```

LOCATION J14

Figure 1.   Game I/O port and statements for an APPLE II.


Analog in:

```
POTX:    X=PEEK(36872)
POTY:    Y=PEEK(36873)
```

1) GAME I/O

Switch in:

```
POKE 37139,0:DD=37154:PA=37137
PB=37152:POKE DD,127
S3=((PEEK(PB)AND 128)=0):POKE DD,255
P=PEEK(PA):S0=((P AND 4)=0)
S1=((P AND 8)=0):S2=((P and 16)=0)
S4=((P AND 32)=0)
```

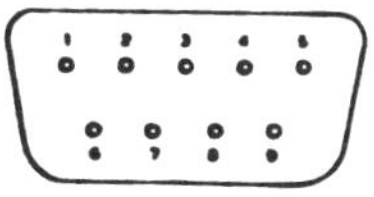

| PIN # | TYPE |
|---|---|
| 1 | JOY0 |
| 2 | JOY1 |
| 3 | JOY2 |
| 4 | JOY3 |
| 5 | POT Y |
| 6 | LIGHT PEN |
| 7 | +5V |
| 8 | GND |
| 9 | POT X |

S0 through S3
are Joy 0
through Joy 3
S4=light pen
or fire button

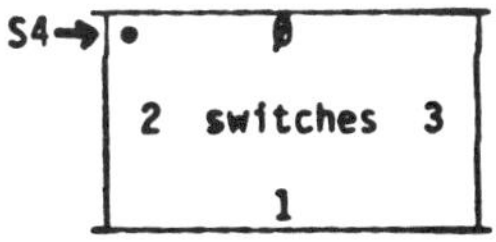

S4→

```
2  switches  3

        1
```

switch arrangement.
in joystick

Figure 2.   Game I/O port of a VIC-20 and necessary statements.

An IBM PC needs the Game I/O board. The pinouts are shown in Figure 3. Programming of the I/O port, for example, for the analog ports is as follows:

```
10 TEMP = STICK(0)
20 XA = STICK(0): XB = STICK(1)
30 XB = STICK(2): YB = STICK(3)
40 PRINT XA,YA,XB,YC
50 GOTO 10
```

These examples show that with minimal programming one can easily construct interfaces to a variety of simple devices. For more complicated instrumentation, analog to digital (A/D) converters are necessary.

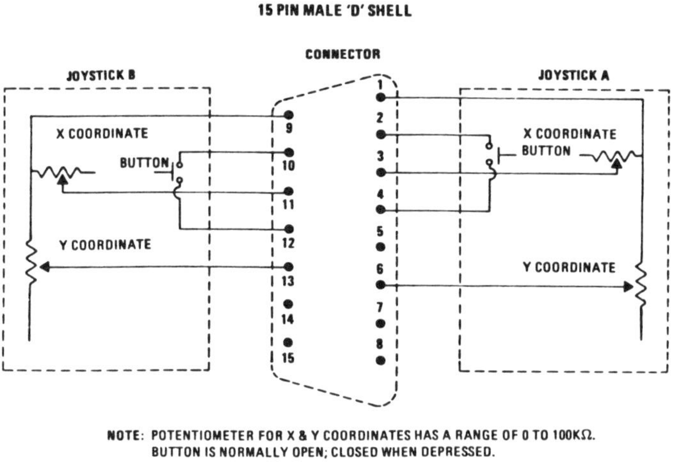
15 PIN MALE 'D' SHELL
CONNECTOR
JOYSTICK B
JOYSTICK A
X COORDINATE
BUTTON
Y COORDINATE
X COORDINATE
BUTTON
Y COORDINATE
1
2
3
4
5
6
7
8
9
10
11
12
13
14
15
NOTE: POTENTIOMETER FOR X & Y COORDINATES HAS A RANGE OF 0 TO 100KΩ.
BUTTON IS NORMALLY OPEN; CLOSED WHEN DEPRESSED.

**Figure 3.** IBM Game I/O connector.

## ANALOG TO DIGITAL CONVERSION

Consider, for example, the conductivity meter shown in Figure 4. The panel meter indicates the conductivity or salinity depending on the switch settings. Figure 5 shows the back (opened) of the instrument. If a voltmeter is attached to the two terminals of the display, voltages will be found to vary between 0 and 50 millivolts. If the instrument is used to monitor the water quality continuously, it is necessary to continuously bring the data to a microcomputer and analyze and record the data. An A/D converter is therefore necessary. There are two considerations in the selection of an A/D converter. First, the speed of the A/D conversion. For many electronic applications several thousand samples per second are necessary. A sampling frequency of 1000 hertz or 1KHz means a thousand samples per second. A/D converters with sampling rates in the thousands of hertz are available for microcomputers, but they need direct access to the system bus of the micro and therefore each is unique to the bus structure of the microcomputer. Data transfer to the microcomputer is in parallel mode.

**Figure 4.** YELLOW SPRINGS conductivity meter.

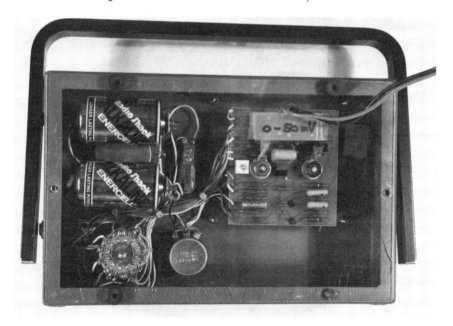

**Figure 5.** Conductivity meter - back opened.

In the water and wastewater field, there are very few applications where such high speeds are necessary. As a matter of fact, sampling every second is too much in most cases. For lower speed of sampling, a serial A/D converter is appropriate. Since the serial transmission of data is standardized, such a device can be attached to any microcomputer which has a serial port and supports the RS-232 protocol. There is a variety of such devices on the market for under $500.

The second consideration is the accuracy of the A/D converter. An 8-bit A/D converter digitizes the analog voltage into 256 (28) individual steps. Thus, if the range of the signal for the A/D converter is 0-5 volts, an 8-bit A/D converter will be able to resolve it to the closest 5000/256 20 millivolts. A 12-bit A/D converter will resolve a range of 0-5 volts to about 1 millivolt (5000/4096 = 1). There are A/D converters with up to 16 bits. Prices for these units increase proportionally to the bit resolution. In most applications a 12-bit A/D conversion is ample, and for many even 8 bits are sufficient.

Figure 6 shows an A/D converter made by Remote Measurements, Inc. of Seattle, Washington. This unit has 16 input channels with 12-bit conversion, 4 digital inputs, and 6 relay outputs. In addition, the BSR controller allows the switching of up to 32 AC sources. The voltage input is expected to range between 0 and 0.4 volts.

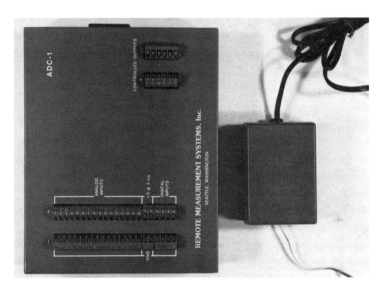

**Figure 6.** Analog/Digital Converter with BSR controller.

## SIGNAL CONDITIONING

If the voltage range of the instrument does not match that of the A/D converter, some signal conditioning is necessary. Several cases may occur:

1.  If the voltage signal is too high, a voltage divider is necessary.
    A sketch of such a circuit is shown in Figure 7.

2.  If the voltage signal is too low, it needs to be amplified.  Figures 8 and 9 are examples of such circuits.
3.  If the input signal is not voltage but current (say, 0-20mA), a small circuit like the one shown in Figure 10 will convert it to voltage.

Again, the programming for a single channel is not too difficult, and a minimal program for a single channel for an APPLE IIC is shown in Figure 11.

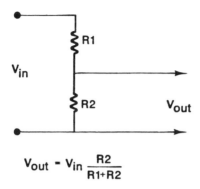

$$V_{out} = V_{in} \frac{R2}{R1+R2}$$

**Figure 7.**  Sketch of a voltage divider.

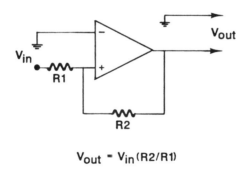

$$V_{out} = V_{in}(R2/R1)$$

**Figure 8.**  Noninverting amplifier schematics

## DATA ACQUISITION

In a conventional water or wastewater treatment plant, many parameters of operation are measured and are usually displayed in a control room on panel meters or strip chart recorders.  To monitor these measurements, it is necessary to find the points behind the panels and recorders where the voltages or currents are present.  Once these points have been found, wires can be attached and the signals can be sent to the A/D converter.  Some signal conditioning may be necessary.  But basically all signals in a plant can be monitored by a microcomputer without affecting the current operation of the plant, and the monitoring can occur concurrently.  The data gathered can be analyzed and stored, and can become inputs to management program.

## CONTROL OF DEVICES

Up to now, only data gathering has been discussed.    To control aspects of operation, say turning on or off a pump, signals must be sent from the microcomputer to the various devices.    The BSR controller shown in Figure 6 (the small box) will send signals over the AC lines and turn on/off a small relay in a module attached to an AC wall outlet (see Figure 12).    The wall module is typically limited to switching a 500 watt load or a 1/3 HP motor.    If heavier units have to be switched, another intermediate relay which carries the appropriate load is necessary.    On many control panels, pushbuttons are present which activate a circuit.    These can certainly be used, and by putting a relay in parallel, the action can be carried out under computer control.

## SUMMARY

There are many aspects of operation of a water or wastewater treatment plant which can be monitored continuously using a microcomputer.  Simple examples have been shown here, and what is needed is to experiment with other possible uses.  The circuits are not difficult and the programming is quite simple.

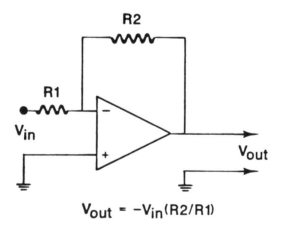

$$V_{out} = -V_{in}(R2/R1)$$

**Figure 9.**  Inverting amplifier schematic.

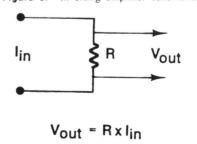

$$V_{out} = R \times I_{in}$$

**Figure 10.**  Current to voltage schematic.

```
  2   S=2                              ;set serial port number - 1=printer
                                                                   2=modem
  3   BA=14                            ;set IIc baud rate - 6=300, 8=1200,
                                                          10=2400, 12=4800, 14=9600
  5   SD=49288 + S*16                  ;Serial Data address
  6   SS=SD + 1                        ;Serial Status address
  7   SA=SD + 2                        ;Serial port parity setting address
  8   SB=SD + 3                        ;Serial port baud rate address
 10   IF S=1 THEN PRINT CHR$(4);"PR#1"    ;initialize printer port
 15   IF S=2 THEN PRINT CHR$(4);"IN#2"    ;initialize modem port
 20   PRINT CHR$(4);"PR#0"             ;return output to screen
 25   BR=BA+144                        ;calc proper byte for baud rate
 26   POKE SB,BR                       ;set serial port to desired baud rate
 27   POKE SA,11                       ;set port for no parity
 30   X=PEEK(SD)                       ;clear port of any old characters
 40   CN=24                            ;set channel for ADC-1 temp sensor
 50   GOSUB 5000                       ;goto channel start/read subroutine
 51   REM                              ;on return will have millivolts*10
 70   DK= Z/10                         ;calc temp in Kelvin = millivolts
 80   DF=(9/5*(DK-273))+32             ;convert to degrees Fairenheit
 90   PRINT "TEMP= ",DF                ;print temperature
100   GOTO 40                          ;repeat forever

5000  POKE SD,CN                       ;select temp sensor; start A/D conversion
5010  GOSUB 5200                       ;wait for ADC-1 response to start command
5020  POKE SD,161                      ;tell ADC-1 to return A/D high byte/status
5030  GOSUB 5200                       ;wait for reply
5040  HB=CH                            ;save high byte from A/D
5050  IF HB >=128  THEN 5020           ;check status for A/D finished
5060  POKE SD,145                      ;command ADC-1 to return A/D low byte
5070  GOSUB 5200                       ;wait for reply and read byte
5080  LB=CH                            ;read low byte from A/D into LBYTE
5085  HM=INT(HB/16)*16:HM=HB-HM    ;mask for 4 high order bits from A/D
5090  Z=LB+256*HM                      ;combine all 12 bits from A/D
5095  SG=INT(HB/16)                    ;select sign bit from A/D high byte
5100  IF INT(SG/2)*2=SG THEN Z=-Z    ;fix sign if negative flag set
5110  RETURN

5200  XS=PEEK(SS)                      ;check status location for incoming byte
5205  SM=INT(XS/8)                     ;select bit specifying byte received
5210  IF INT(SM/2)*2=SM  THEN 5200    ;if no byte received, then try again
5220  CH=PEEK(SD)                      ;get character from received data buffer
5230  RETURN                           ;return to calling program
```

**Figure 11.** Listing of program for Apple II and one analog input.

**Figure 12.** BSR wall switch.

# CHAPTER 5

## USING COMPUTERS FOR PROCESS CONTROL AT SMALL TREATMENT PLANTS

Phillip Loud, P.E.
Ayres, Lewis, Norris, & May, Inc.
Ann Arbor, Michigan

## INTRODUCTION

With the introduction of personal computers to the water and wastewater plant environment, those smaller plants heretofore unable to afford computer systems are now able to take advantage of these valuable tools.

Prior to the advent of the microcomputer, those systems installed in municipal facilities were commonly equipped with customized sofware specifically designed for that facility and its operation. These systems often were accompanied by a large price tag. The popularity of personal microcomputers on the other hand has opened up a competitive market for generic software which can be used by different types of facilities to improve recordkeeping and operational efficiency.

This chapter addresses those readily avialable types of software applicable to process analysis and control of small water and wastewater facilities. Emphasis is placed on utilization of software which is relatively inexpensive and "user friendly."

There are many applications in treatment plant process analysis for the use of standardized software products. These products generally include graphics programs, spreadsheets, data base management, word processing and communication software. Typical applications include data collection, trend analysis, process optimization, report generation, training and remote data base acquisition.

The use of a microcomputer can be initiated by the staff of small facilities without the immediate need for custom software or expensive hardware capabilities.

## STANDARDIZED SOFTWARE PRODUCTS

### Graphics

As stated above, the five most commonly utilized standard software packages suitable for plant operations include graphics, spreadsheet data base management, word processing and communications.

Graphics programs are designed to provide the user with an opportunity to visually examine the characteristics of interrelated data. The basic purpose of such software is to present quantitative information in the form of images or pictures such as bar line or pie charts. The output from such software can be simply viewed on the computer monitor or transmitted to a printer or plotter for hard copy.

Most graphics programs allow for the manipulation of scale, x and y legends, and the use of multiple variables. The charges can be stored on "floppy" disks for future reference or updating. Stored data can be changed periodically to include more recent plant process data. Graphing is commonly used to represent process trends, comparative quantities or the interrelationship of process variables.

## Spreadsheets

Spreadsheet programs or electronic worksheets are simply designed for the organization and manipulation of data in a large matrix format. The basic purpose of a spreadsheet is to provide the user with a grid of rows and columns in which text, numeric data, or formulas can be initially stored. The information can then be manipulated through the use of a wide range of mathematic functions available to the user. Typical mathematical functions would include totals, averages, maximum, minimum, and complex formulations, including addition, subtraction, multiplication, division, etc.

Typical application of spreadsheets includes budgeting, inventory, daily/weekly/monthly process summaries and financial analysis. The operator of these programs has wide flexibility over the structure of each row or column in order to meet specific needs.

## Data Base Management

Data base management programs provide a system for the filing and retrieval of information. The purpose of these programs is to provide a system for organizing, storing, retrieving, and modifying a specific data base. These programs are often referred to as electric filing cabinets.

Typical features of a data base management system are the ability to define a page format for the file; enter, change or delete records; and retrieve selected records given simple identification criteria. These systems can often be augmented by report generators which will tabulate or summarize specific items within the data base. In fact, simple mathematical manipulation is often possible, including provisions for the calculation of totals, averages and sorting by size.

Data base management systems are often best suited to establishing equipment inventories, preventive maintenance records and manpower scheduling.

## Word Processing

Word processing software is designed for the creation and manipulation of text. These systems provide an opportunity for the storage and later editing or modification of information commonly in text form.

Typical features include text manipulation, underlining, removal, labeling, appending, duplication, bold face, etc. Word processing allows plant management personnel, for example, to generate memos, letters, reports, contracts and training documentation without requiring secretarial assistance and allowing for efficient text changing.

## Telecommunications

Telecommunication programs are designed to link computer systems via conventional telephone lines. The basic purpose is often to allow remote microcomputers to communicate with larger "host" computers for the purpose of transmitting or receiving data.

Typical application of telecommunications software is for the access of information from service bureaus (i.e., stock quotes, electronic shopping, technical information sources) or terminal emulation for a larger mainframe computer.

## COMPUTERIZED PROCESS ANALYSIS

### Data Collection and Generation

The collection of pertinent operational data from a water and wastewater plan can be greatly facilitated through the use of a microcomputer. Data can be input to the computer either manually or through interface with analog or digital sources. The use of readily available generic software is best suited to the manual input of plant data.

Raw data sources in a plant environment are limitless and many are currently hand tabulated in order to evaluate process activity or generate reports. Examples of these sources include BOD, suspended solids concentration, phosphorus loading, dissolved oxygen concentrations, air usage, flow, pH, chemical usage, energy consumption, etc.

Many plants currently utilizing microcomputers will input plant data to a spreadsheet program so that the data can be mathematically manipulated. For example, in a wastewater plant given influent flow in million gallons per day (mgd) and BOD concentration in milligrams per liter (mg/l), the spreadsheet can generate a column of BOD loading in pounds per day with a simple key entry. Average BOD concentration can be calculated along with total BOD loading for a month. The flexibility is the option of the user of the microcomputer. Table I presents a spreadsheet summary of certain components of the operation of an activated sludge wastewater treatment system. Basic input data include month, flow, influent and effluent BOD and air volume fed. The total pounds of BOD removed was calculated, as was the ratio of cubic feet of air per pound BOD removed, which is a common indicator of excess air usage. The last column presents a typical value for the ratio based on very simplistic summer and winter operating guidelines. It can be readily seen that the plant appears to be applying too much air during many months of the year.

Data generated by a spreadsheet program working with data input by the operator can be used as input to a graphics display program for process trend analysis. The same kind of data presented in Table I can be graphically displayed concurrently with

**Table 1.    Activated Sludge System Operating Summary**

CITY WASTEWATER PLANT
AUDIT YEAR 1985

| MONTH | AVE. FLOW (MGD) | INFL. BOD (MG/L) | EFFL. BOD (MG/L) | LBS. BOD REMOVED | CUBIC FT AIR (*1000) | CF AIR PER LB BOD REM. | TYPICAL VALUE CF/LB |
|---|---|---|---|---|---|---|---|
| JAN | 3.2 | 238 | 3 | 6271.7 | 6589.8 | 1050.7 | 900 |
| FEB | 4.77 | 171 | 3 | 6683.3 | 9299.1 | 1391.4 | 900 |
| MAR | 5.15 | 201 | 4 | 8461.3 | 11395.4 | 1346.8 | 900 |
| APR | 4.12 | 221 | 3 | 7490.7 | 14225.3 | 1899.1 | 1200 |
| MAY | 3.17 | 235 | 2 | 6160.0 | 11011.8 | 1787.6 | 1200 |
| JUN | 3.41 | 163 | 2 | 4578.7 | 5497.2 | 1200.6 | 1200 |
| JUL | 3.42 | 182 | 2 | 5134.1 | 5657.3 | 1101.9 | 1200 |
| AUG | 3.68 | 196 | 2 | 5954.1 | 7723.3 | 1297.1 | 1200 |
| SEP | 4.9 | 143 | 2 | 5762.1 | 5752.0 | 998.2 | 1200 |
| OCT | 4.07 | 197 | 2 | 6619.0 | 5959.8 | 900.4 | 900 |
| NOV | 4.53 | 178 | 2 | 6649.3 | 6779.9 | 1019.6 | 900 |
| DEC | 5.57 | 160 | 2 | 7339.7 | 8799.6 | 1198.9 | 900 |

other operating data such as in Figure 1.    Here the activated sludge system is represented by three basic parameters: air usage, dissolved oxygen concentration, and mixed liquor suspended solids.    The interaction of these parameters is clearly visible.

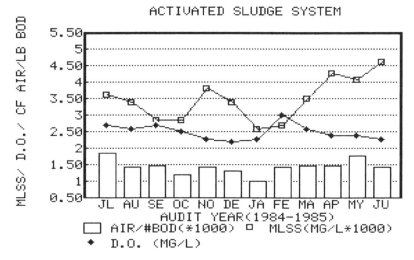

**Figure 1.**    Activated Sludge Operating Summary

Another example of utilizing the spreadsheet program and graphics together to display system operating characteristics includes examination of pump performance.

Table 2 presents a summary of the operating points of a wastewater pump over a range of total heads.    Input data included head, flow and efficiency.    Brake horsepower was calculated by the spreadsheet program by introducing the formula

$$BHP = QH(3960 \times Ep)$$

where

$Q$ = flow in gpm,
$H$ = head in feet, and
$Ep$ = pump efficiency

With a simple keystroke, the entire column of BHP's was generated.    The data from Table 2 could then be graphically displayed to give the operator a clear understanding of the impact of varying head conditions on pump output capacity and efficiency.

**Table 2.**  Pump Performance Data

| TOTAL HEAD (FT) | OUTPUT FLOW (GPM) | PUMP EFFIC. (%) | BRAKE HORSPWR (BHP) |
|---|---|---|---|
| 46 | 2000 | 46 | 50.5 |
| 45 | 2500 | 58 | 49.0 |
| 44 | 3300 | 70 | 52.4 |
| 41.5 | 4200 | 79 | 55.7 |
| 39.5 | 4800 | 83 | 57.7 |
| 38 | 5400 | 84 | 61.7 |
| 36 | 5800 | 84.5 | 62.4 |
| 34 | 6200 | 84.5 | 63.0 |
| 32 | 6600 | 84 | 63.5 |
| 29 | 7000 | 82.5 | 62.1 |
| 27.5 | 7200 | 80 | 62.5 |
| 26 | 7500 | 70 | 70.3 |

Data collection need not be limited to process related information but can include equipment data, maintenance activities and inventory control.    For these functions, spreadsheets and data base management or filing programs are often used because of their ability to be readily updated and sorted.    Table 3 presents a summary of the pump and motor nameplate data recorded for two separate water plants.    The information can be readily updated and permanently stored for reference on floppy disks or on a hard disk.

Table 3.  Water Plant Pump/Motor Nameplate Data

| STATION/PLANT | PUMP NO. | HP | MOTOR NAMEPLATE DATA | | | PUMP NAMEPLATE DATA | | |
|---|---|---|---|---|---|---|---|---|
| | | | VOLTAGE | F.L.A. | RPM | TDH | GPM | MGD |
| LAKE HURON | RW1 | 1250 | 4600 | | 450 | 53 | 69444 | 100 |
| | RW2 | 1250 | 4600 | | 450 | 53 | 69444 | 100 |
| | RW3 | 2250 | 4600 | | 327 | 53 | 138888 | 200 |
| | RW4 | 2250 | 4600 | | 327 | 53 | 138888 | 200 |
| | HL1 | 5500 | 13800 | | 600 | 435 | 41700 | 60 |
| | HL2 | 5500 | 13800 | | 600 | 435 | 41700 | 60 |
| | HL3 | 5500 | 13800 | | 600 | 435 | 41700 | 60 |
| | HL4 | 5500 | 13800 | | 600 | 435 | 41700 | 60 |
| | HL5 | 5500 | 13800 | | 600 | 435 | 41700 | 60 |
| NORTHEAST | RW1 | 1250 | 4600 | 245 | 400 | 80 | 41600 | 60 |
| | RW2 | 900 | 4600 | 177.2 | 450 | 80 | 34750 | 50 |
| | RW3 | 1250 | 4600 | 177.2 | 400 | 80 | 41600 | 60 |
| | RW4 | 900 | 4600 | 177.2 | 450 | 80 | 34750 | 50 |
| | RW5 | 1250 | 4600 | 177.2 | 400 | 80 | 41600 | 60 |
| | RW6 | 900 | 4600 | | 450 | 80 | 34750 | 50 |
| | RWD7 | 200 | 2300 | 48.2 | 885 | 80 | 5200 | 7.5 |
| | RWD8 | 200 | 2300 | 48.2 | 885 | 80 | 5200 | 7.5 |
| | HL9 | 3000 | 2300 | | 514 | 241 | 36111 | 52 |
| | HL10 | 3000 | 2300 | | 514 | 241 | 36111 | 52 |
| | HL11 | 1750 | 2300 | 514 | | 200 | 34006 | 49 |
| | HL12 | 1750 | 2300 | 514 | | 200 | 34006 | 49 |
| | HL13 | 1750 | 2300 | 514 | | 200 | 34006 | 49 |
| | HL14 | 1750 | 2300 | 510 | | 200 | 34006 | 49 |
| | HL15 | 3000 | 2300 | | 514 | 200 | 36111 | 52 |
| | HL16 | 3000 | 2300 | | 514 | 200 | 36111 | 52 |
| | HL17 | 3000 | 2300 | | 514 | 200 | 36111 | 52 |
| | HL18 | 1750 | 2300 | 510 | | 200 | 34006 | 49 |
| | HL19 | 1750 | 2300 | 514 | | 200 | 34006 | 49 |
| | HL20 | 1750 | 2300 | 514 | | 200 | 34006 | 49 |

## Process Modeling and Analysis

Once the proper operational data has been assembled, whether on a spreadsheet, in graphic form, or in the file of a data base management system, the information is available for use in process modeling and analysis.    Process modeling is often performed through the use of time based graphic trend analysis.    Data collected over time can be displayed graphically and the change in various system parameters can be examined concurrently.    The interaction of many process variables is often best viewed in graphic form then related back to the tabulation generated by the spreadsheet program.

## Process Control

Process data generated and later analyzed can be used to initiate operational or maintenance changes which result in improved effluent quality, reduced plant upsets, increased process and energy efficiency or creation of more available staff time for other projects.    An example of this is presented in Figure 2 where four high service water pumps of identical size have been tested for overall efficiency (wire to water) over a range of pressures and the various efficiencies plotted.    This graph in turn can be used by plant operators to dictate which pump is most efficient at any system pressure.    Obviously, at pressures of 80 to 85 psi, for example, high lift pump number 19 would be preferred as lead pump under ideal conditions.

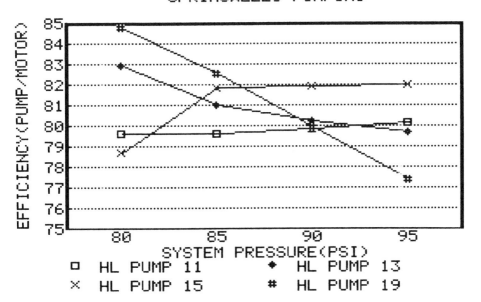

Figure 2.   Springwells High Lift Pump Efficiencies

With many microcomputers, process control can be designed to evolve from an initial program of total manual implementation to partial remote keyboard system control and on to closed loop control.    In many cases, simply using the knowledge gained in process analysis can be incorprated into process changes implemented through existing systems controls.    If desired, the microcomputer can be used as a central control point to send out analog or digital control signals to various process units. Ultimately the user can develop mathematical algorithms to examine incoming process data and generate automatic process control signaling based on analysis built into the algorithms.

## CONCLUSION

It is important in the selection of computer system hardware to be foresighted enough to anticipate the ultimate needs.    A microcomputer capable of closed loop. automatic control can be purchased initially with the capability of being expanded to provide that type of control.    Not all microcomputers have this flexibility or expandability.    Most importantly, it should be recognized that in many plant applications, a microcomputer with basic peripherals (printer, disk drives, monitor) can be coupled with readily available generic software to allow a plant to begin computerization of a number of activities without a large expenditure.

# CHAPTER 6

## USING COMPUTERS FOR PROCESS CONTROL AT LARGE TREATMENT PLANTS

David C. Mohler, PE
McNamee, Porter & Seeley
Ann Arbor, Michigan

## INTRODUCTION

Certain basic functions must be carried out in any design regardless of how a control system is implemented. These include the following:

1. Measurement
2. Process control
3. Operator information
4. Management information

Analog systems and computer-based systems carry out the same functions though the methods employed differ. Indeed, the analog instrument-based systems involve the use of analog computers (controllers) to solve the valve positioning problem. The digital process computer does exactly the same thing, but the problem is solved numerically.

The analog approach has an advantage in that the failure of a controller usually affects only the control loop in which it is involved. A cascade effect, however, may allow the failure of a critical controller at the head of a plant to adversely affect processes downstream. The typical process computer handles from eight to several dozen control loops; thus, the failure of a process computer has a more dramatic effect. However, current computer technology renders such machines extremely reliable, especially when dual redundant designs are employed. Fault tolerant and continuous uptime machines are becoming common in the process control industry.

## ANALOG CONTROL

Methods employed to achieve the above functions vary depending on the implementation of the control system. Any modern system will accomplish the first two functions automatically whether it is built using analog devices or process control computers. The second two functions depend largely on manual methods when analog instruments are used. For instance, if the operator needs to know what devices are signaling alarms around the plant, it is necessary to visit the various control panels to visually determine this information. If the plant manager wishes to have a summary of the previous day's operations, he or she must depend on the observations of the staff. It is common to assign staff to write down totalizer readings on a tally sheet each day.

## COMPUTER CONTROL

With computer control, the above information is always at the fingertips of plant operators and managers. Many operators have planimetered a recorder chart in order to analyze the information. With computer technology the process computers can deliver process information to a spreadsheet program or statistical analyzer for on-the-spot analysis.

Alarm reporting is centralized, with the added benefit of a printed history of alarm events for the plant. Analysis of the sequence of unusual plant events often leads to a quick identification of failed devices in the plant. Other information is brought to the plant operator's fingertips by the computer. This includes measurements from any process instrument such as flows, levels, dissolved oxygen, chlorine residual and pH. Data about the machinery of the plant, such as run-times and number of operations, can be gathered for analysis or for use by maintenance management software.

Two strategies may be used for analytical and planning tasks such as maintenance management and monthly operating reports. First, these tasks may be done by the process control computing system. This has been the approach of choice until recently. Now, such tasks often can be accomplished more conveniently with desktop computers. In this case, the control computers act as data-gathering devices that hold the process information until it can be transferred. This allows the use of commercially available desktop computer software for maintenance management and report preparation.

Process computers provide enhancements to measurement and control functions. In process measurements, the acquired signals can be made more useful by employing conditioning algorithms that mask the noise, or jitter, in signals delivered by some instruments. The algorithms will notify the operator when the signal has gone out of range, indicating that the sensor has failed or lost power. Information can be obtained from signals that are not easily developed with analog instruments. For instance, an alarm dependent on the rate of signal change can be useful. This is difficult with analog devices, but requires only a few lines of programming with the computer.

The mathematical precision of the computer allows for more accurate modeling of the process being controlled. Complex continuous simulations can be carried out that result in optimized plant operations. For instance, chemical dosages for phosphorus removal can be held at the minimum required to meet the discharge requirements. By supplying the process computer with data from the laboratory and using it together with live process data, the process control can be refined quickly to achieve a minimum operating cost.

## KINDS OF PROCESS COMPUTER SYSTEMS

Computerized control systems come in a variety of arrangements, depending on the needs of the plant and the process to be controlled. Since the first such systems were installed in the late 1960's, system arrangements have gone through an evolutionary process.

## Early Systems

Early wastewater and water process systems of the late 1960's and 1970's were of three general types.

1. Direct Digital Control (DDC)
2. Digital Supervised Control (DSC)
3. Digital Monitored Analog Control (DMAC)

These general types usually employed the assistance of conventional analog instrumentation to enhance the reliability of the system.     Computers of that time were not nearly as reliable as computers today.     This is true in both a hardware and software sense.

Direct digital control systems of early design (see Figure 1) can be characterized as follows:

.   control computer(s)

.   redundant control computers

.   all incoming and outgoing signals wired centrally direct to the control computer

.   direct computer control of final elements

.   custom, one-of-a kind, process control software

.   low-level assembly language programming

.   process changes required considerable programming effort to accommodate

.   closed designs such that information could not easily be imported or exported from the control computer

.   analog backup controls - this usually made the installation quite expensive because of the redundant hardware

.   computer failures could cause serious disruption in operations

.   small disk storage in the range of 0.5 megabytes (MB) to 10 MB

.   no tape or off-line storage for backup or archiving

.   consoles were very simple without video displays

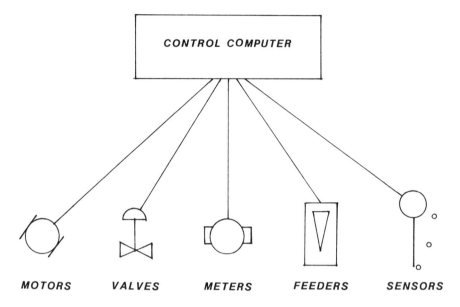

**Figure 1.** Direct wire systems

Supervisory computer control systems are in many ways similar to the above, but are different in the following ways:

- computers would generate set points for use by conventional analog controllers

- computer failures have a lesser impact on operations

With digital monitored analog control, the central computing facility serves only as a data gatherer for plant operators and managers. The actual control in this kind of early system is carried out by conventionally wired analog instruments and relay logic. Such designs proved to be expensive because a conventional control system was being bought in addition to the computing system.

## Modern Systems

Systems of more recent vintage in wastewater and water treatment and collection systems (see Figure 2) have several attributes in common. These include:

- distributed control where a multitude of control computers are used throughout a plant to control processes nearby

- high speed local area networks (LANs) allowing for communication between the process computers

- use of inexpensive computers for operator interfaces

- lack of control panels

- hardware of exceptionally high reliability

- exposure to failures can be limited to small portions of the process

- easy to program using process control languages

- large amounts of on-line storage are typical, from 20 MB to 100s of MB

- data archiving and backup through the use of high capacity tape cartridges

- sophisticated operator interfaces to the process using consoles and network monitors

- enhanced operation of the control center through the use of robust support programs

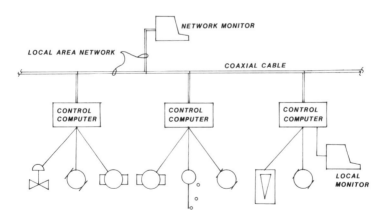

**Figure 2.** Local area networks.

## USER INTERFACES

A user interface is the means by which a person is able to communicate his or her desires to the process. Traditionally, control panels have been used to provide the interface. In more recent times, the use of video display terminals and small computers has become more commonplace.

## Control Panels

Control and instrument panels represent the traditional approach for an opeerator to obtain process information and to communicate to the process controls the changes necessary in operations.    Such panels consist of an array of electro-mechanical devices such as indicators, push buttons, lights, switches, relays and analog control instruments.

These designs involve the use of large spaces that might better be used for other purposes. Because of the basic mechanical nature of control panels, reliability can be a problem.    This is particularly true if the equipment is in a hostile environment.

Control panels are fixed in their layout due to the fact that instruments and devices are mounted in holes cut in the metal face of the panel.    Panel wiring is also fixed according to the design of the plant's process.    This presents a problem if the process were to change or the plant were expanded.    The renovation cost can be quite large.

## Control Console

Control consoles have been part of most process computer control system designs since the late 1960's.    In the typical console, several video display terminal (VDT) screens present process information to the plant operator.    Video displays are high resolution, 480 lines by 800 pixels, or 48 lines with 80 characters per line.    these displays usually employ several colors to highlight important information.    Up to eight colors are used to enhance the operator's ability to recognize important information.    For instance, process flow lines might be aqua, flow values in green if normal but red if abnormal, and so on.

The following kinds of displays and printed reports can be found in a typical treatment system:

- alarm summaries

- process flow diagrams with live instrument readings

- trend plots

- process indicators and controllers allowing the operator to supply set points and outputs to valves

- operating reports

- machine run times

- programming facilities

- System security to limit changes to critical parameters by only those authorized

Typically, consoles are custom fabricated to the specific plant needs.    In addition to the VDTs, there are hardcopy devices that can record plant measurements, alarm messages and reports.    Associated with each VDT is a keyboard that the operator uses to instruct the computer. Instructions might be to call up a new display, to print a report, to review operations or to change a process parameter.

Control consoles (see Figure 3) present their own class of maintenance problems to plant management because of dependency on large quantities of electronic parts. Console systems usually require the support of a nearby data gathering computer to handle communication from the operator to the outlying control components. These computers represent a significant investment and require that conventional data processing methods be used to manage the computer system. Representative computers are Digital's VAX-11, PDP-11 and Hewlett-Packard's HP-1000.

Tasks that are not strictly process control tasks can be accommodated by this class of computer. These include maintenance management and scheduling, inventory control, machine run-time acquisition, laboratory data input and monthly operating report preparation.

A new area that we believe is important to plant operators and managers is the ability to include instructions to the operations staff on-line. It is now easy, with the power of today's computers, to implement a plant's operation and maintenance manuals for display on the console VDTs. In this use of the computer, a user would ask the machine for help with a particular topic. That person might enter something like "Help SUMP-PUMPS," and the computer would respond with a description of the plant's sump pumping operation. Then it would list the available subtopics of SUMP-PUMPS and ask the user for the subtopic to be shown. This process continues until the operator obtains the level of detail necessary to answer his or her question.

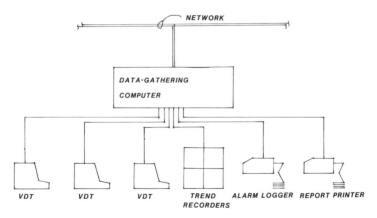

**Figure 3.**  Typical control console.

## Network Monitors

A concept using a device known as the network monitor has become popular in recent years. The network monitor is a specially prepared desktop microcomputer such as the IBM-PC. A standard IBM-PC/XT or IBM-PC/AT has several components added in order for it to be an effective monitoring tool.    Included are special communications hardware and software allowing it to talk directly to the process control computers.    A high resolution graphics adaptor card is added allowing for color displays of 400 lines by 640 dots.

Desktop computers provide virtually the same functions as the more costly console approach with only a slight loss in computational ability.    An advantage of this approach is that most popular commercially available software can be used concurrent with process monitoring activities.    Users have the benefits of controlling processes, gathering data and being able to analyze the data using their favorite spreadsheet or data base software.    A special operating system is used that allows the machine to switch from one task to another rapidly in rotation giving the impression that the computer is doing many things at once.    This is called multitasking and is brought about through the use of an operating system called "Concurrent CP/M."    While the computer continually accepts data from the process, it also allows one to use standard MS-DOS programs such as spreadsheets, word processors and data base management systems.

Many designs use several network monitors, allowing for increased reliability and the ability for several operators to monitor different parts of the process at the same time.    This approach allows monitors to be placed conveniently in process areas to replace control panels. Industrial versions of the IBM-PC are used where the environment is less than ideal.

## CURRENT TECHNOLOGY

This section reviews the current state of computerized control technology as applied to water and wastewater systems.

## System Components

Computer control systems consist of a collection of componenets each having a specific purpose.    These include process input and output, control computers, communications networks and user interface systems.

### Control Computers

Programmable process controllers (PPC) are computing elements that are generally strong in analog process control.    All the necessary elements are built-in to simulate all kinds of traditional analog instruments including controllers, ratio stations, trip switches and function generators.    They have evolved out of the process control industry which has in the past built control panels and instruments. Also included are the essential ingredients for sequential control of pumps and motors.

PPCs have a very good record for reliability.    For critical applications such as sewage pumping, the controller can be made more reliable by including a standby process computer at the same site.    In this case, one machine watches the other, making sure that the computer in control is actually carrying out its tasks.    If a failure should occur, the standby computer will take over control and send an alarm message to the control center.

Programmable logic controllers (PLC), on the other hand, are strong at sequential logic and contact input and output. PLCs have evolved out of the need for computerized machine tools and are relatively weak in analog control. They are capable of decision logic, sequential control and timing functions. If any analog ability is included in a particular PLC, it is very rudimentary and difficult to use.

## Networks

Virtually all computerized control systems today employ a distributed approach wherein direct digital control devices are placed at strategic locations around the plant for the control of nearby processes. This concept is similar to the more traditional method, in which several control and instrument panels would be used in the outlying locations of the plant. Conceptually, the control designs are similar. Each control location is responsible for the process local to it.

Communications networks are the backbone of modern computer-based control systems, providing the means by which the process control computers, network monitors and control console exchange information. Physically, the network consists of communications circuit boards at each location that are interconnected by means of a simple cable. This cable is a coaxial cable similar to a CATV cable, a fiber optic cable, or, in some cases, a twisted pair of wires. Such a cabling system is used in place of the many control and signal wires that are run throughout a plant in analog control designs. Plant wiring costs and maintenance problems can be further reduced when the system design calls for the installation of input and output hardware directly in motor control centers.

Most networks today are arranged in a loop or daisy-chain fashion in which a cable is run from point to point around the plant. Adding a new point, or node, merely involves injecting the new node into the loop and maintaining the software to recognize the new controller.

All networks in common use today employ high transmission speeds on the order of one million bits per second or faster. To enhance reliability, standard protocols are used which have been developed by the IEEE. Two of the more common approaches are Ethernet and the Manufacturing Applications Protocol. Each features error detecting and correcting techniques.

To further increase the reliability of the network, it is common to provide all components in duplicate, that is, two sets of communications cards and two sets of separately routed cables.

## Control Centers

The result of all the above is that now for the first time it is possible to seriously consider control system designs that have no control panels. Recall that purposes of the control panel are as follows:

1. house the control logic

2. provide a place to mount instruments and analog controllers

3. be a location for push buttons, lights and switches

4. house recorders

5. be the operator's interface to the process

These functions are provided by other means in the computerized control systems.

1. The PPC and PLC house the control logic.

2. The PPC and PLC simulate the action of various instruments and controllers.

3. The network monitor gives an equivalent function of the push buttons, lights, etc.

4. The network monitor and/or data-gathering computer records process measurements.

5. The network monitor is the operator's interface to the process.

## PROGRAMMING

Programming of modern process control computers is a straightforward, simple task that can be done by anyone with a minimum of training. It is more important today to be knowledgeable about the process to be controlled than to be a "wizard" at programming computers. Compare this to the situation of early systems.

In earlier times, the main problem in establishing a computer control system was the techniques of programming the computer efficiently so that the programs would run fast enough to actually be able to control the process. As mentioned above, each early system was unique and, in many cases, proved extremely difficult to maintain or modify when the need arose.

Today, we have computers and programming systems that are much more robust. We can program the system without disturbing on-going processes using standard control program modules that allow us to define the proper control methods by "wiring" software modules together to form a complete system. A newly created control program can be down-loaded to a PPC over the network for testing. If it works, the program is made permanent; if there are problems, the original control program can be quickly restored.

## Wiring With Software

Programming of modern process control computers is a relatively easy process that is governed by logic. One does not need to be concerned with questions of how the computer acquires or generates signals, or the sequencing of tasks.

Process computer programming tasks can be broken down into the following: (1) identify each instrument signal coming into or out of the system. (2) identify and connect together standard control modules. (3) identify each point on the communications network, and (4) lay out displays and printouts.

1. Identify each input and output signal

- physical wiring terminals

- assigning a name to the signal for all other references

- sampling ranges and units

- engineering ranges and units

- alarm limits in real world units

2. Identify and interconnect the software control modules

   - averagers - accept two or more signals for averaging their values

   - calculations - combine several signals to give new values necessary for control. These are programmed as algebraic expressions like those used in BASIC or Fortran.

   - comparators - compare two signals and tell which is larger

   - controllers - simulate all kinds of process controllers including ratio stations, proportional-only, non-linear, and proportional plus integral plus derivative

   - decision logic - important for sequential control of pumps and blowers

   - differentiators - determine the rate of signal change

   - event counters

   - function generators - important for implementing optimal control strategies. Useful for selecting the most efficient pump turn-on and turn-off points for instance.

   - integrators - typically flow or power totalization

   - lead/lag

   - loggers - give messages about alarms and other process events

   - peak detectors - give the highest instrument reading until reset

   - pulse counters - useful in some telemetry applications

   - timing modules

   - sequencers - give step-by-step operations for tasks such as filter washing

3. Identify modules for communications with other process control units on the local area network

4. Lay out displays and printouts

- simulate analog indicators (horizontal and vertical bar graphs) using color

- trending

- alarm messages

- flow diagram and other video displays; again color is very important

- printed reports

The major equipment manufacturers provide a comprehensive set of tools with their systems to develop the process control programs. These tools are inexpensive compared to the total system cost. For an IBM-PC being used as a network monitor, the development tools cost from $5,000 to $10,000, depending on the supplier and configuration. For a larger computer such as a VAX-11, the cost is $20,000 or more.

## Example Control Loop

A typical control loop that might be found in an activated sludge sewage treatment plant is shown in Figure 4. In this example, the return activated sludge flow (FE-2) is to be controlled in proportion to the plant influent flow (FE-1). The percent of return in proportion to the sewage is set on a ratio station (FF-1) that computes the desired amount of sludge flow. This is the set point for the flow controller (FC-2), which in turn compares the set point to the actual flow and modulates the throttling valve as needed.

In programming the loop, we first define the signal FE-1 by telling the computer which terminals the signal can be acquired from, the engineering units (gpm or mgd), the signal zero value (usually 0.0) and the signal span or range value. In a like fashion we define signal FE-2. The set point for FC-2 is calculated with a proportional-only controller labeled FF-2. We set up this controller to act as a ratio station, specifying the input source (FE-2) and naming the output signal (FE-2SP). The ratio or percent of return in this case is manually set by the plant operator and would be supplied through the use of a video display.

We can now define the flow controller FC-2 by supplying the input signal name (FE-2), the set point signal name (FE-2SP) and an output signal name (FE-2OP) that represents the position the valve is to assume. Additional information in the form of the tuning parameters for the controller are supplied and the controller is complete.

The final task necessary to complete the control loop is to deliver the signal FE-2OP to wiring terminals so that the physical valve can receive the commands of the computer. All that is necessary is to supply the name of the output (FE-2OP), the units (percent valve open), the zero value (0.0%) and the range (100.0%). At this point the loop programming is complete and ready for testing.

## Cost of Programming

Traditionally, process computer systems are programmed by the vendor. This provides the advantage of having programming done by people who are experts in the process

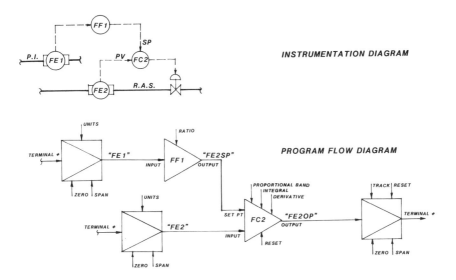

**Figure 4.** Typical programmed control loop.

control language of the system.   The disadvantages are (1) the programmers are not experts in the process to be controlled, and (2) the cost of such programming represents a major portion of the system cost. Many well-designed computer control systems fail to meet the needs of the plant because vendor programmers do not understand the fine details of the process to be controlled.

In the last few years, we have observed a trend away from vendor programming.  Owners of such systems are insisting that they be able to maintain and modify the process control programs with their own staff once the system has been completed.   We feel this attitude is correct for the following reasons.

   •   outside programming from vendors is expensive ($100 per hour
       and more)

   •   it is difficult to communicate your exact needs to someone that
       doesn't know the process intimately

To answer the above need, several things can be done.   First, avail yourself of any training opportunities.   Insist that the system construction contract include adequate training. Second, consider not having the equipment vendor do the initial programming; do it yourself or have the consultant that designed your process do it if qualified. Third, make sure that your staff is involved in the start-up and check processes.

## CONCLUSION

Computerized process control systems are the current technology.  Systems using computers to run wastewater and water treatment plants have been in use since the late 1960's.   During the intervening years, the technology has matured into reliable systems that are inexpensive to own and operate.

Computerization brings several clear benefits to the management of wastewater and water treatment plants:

- changes in process technology are easy to accommodate

- plant wiring costs can be significantly reduced during construction

- large fixed-function control panels can be avoided

- process information is quickly available to operators and managers

- plant operating cost can be minimized through the use of more sophisticated process models

# Chapter 7

# Automation of the Water and

# Sewer Billing Process

Robert McVay, P.E., and Charles Secrest
Genesee County Water and Waste Services
Flint, Michigan

## INTRODUCTION

Medium sized utility offices have generally required the speed and memory capacity of a mainframe computer for processing data, printing of bills and report generation. These machines are highly complex, large and expensive to operate, and require significant space and climate control for proper operation. Operation of these machines is limited to those with extensive training in computer science. Mainframe computers must be supported with specific software which is usually leased at significant cost. Little flexibility is provided by these machines for producing custom reports.

Recent advances in computer technology have resulted in the production of minicomputers with capabilities and computing power now approaching that of larger machines. Minicomputers offer substantial savings in maintenance and support costs. When these machines are used in combination with microcomputers, considerable flexibility for custom report generation is provided and improved access to the the minicomputer is retained.

Technological advances have additionally resulted in the production of solid state interrogators for meter reading purposes. These "smart guns" are actually microcomputers which store data in a format accessible to larger machines.

This chapter demonstrates the cost savings and efficiencies that may be realized by choosing an automated billing system, maximizing the benefits obtainable from a combination system of mini- and microcomputers and solid state meter reading devices. The application of this equipment to the billing process and working examples of the varied uses of the computer combination will be presented.

The equipment is utilized at the Genesee County Drain Commissioner's Division of Water and Waste Services, a medium sized water and waste utility in Flint, Michigan.

67

## GENERAL DESCRIPTION OF UTILITY

Water is provided by the Detroit Water and Sewerage Department to the city of Flint for wholesale to the Genesee County Drain Commissioner's Division of Water and Waste Services. The Division, acting in the capacity of county agent, provides retail and/or wholesale service to ten townships and six cities through a system of individual and master meters.

Approximately 16,000 water customers are serviced. About 7,000 of these are direct customers, with the county providing meter reading, billing, file updating and maintenance services. A portion of these services is provided to wholesale customers.

The county sewerage system consists of collection, pumping and sewage treatment for approximately 55,000 customers. Seventeen townships and nine cities are serviced by the county system. Utility operation, maintenance and billing functions are provided by the division.

Staff functions are performed by data processing, accounting, and operations and maintenance departments. Engineering support is provided by a design and construction section.

The Division employs approximately 100 people, including 54 unionized personnel. Professional staff includes engineers, computer programmers, accountants and various licensed water and wastewater operators.

The Division recently abandoned its mainframe computer and now utilizes a combination mini- and microcomputer system.

As part of the billing system revamping, all of the county's 7,000 water customers are being transferred to the direct read-bill system. At present about 3,000 customers have been transferred.


## EQUIPMENT CONSIDERATIONS

In early 1984, the Division's Data Processing Department examined the feasibility of downsizing the existing mainframe (IBM-4331) computer and replacing it with an IBM 36 microcomputer.

The downsizing was undertaken due to cost considerations, including the cost of maintenance, leased software and software support or upgrading of the old machine. Although memory and speed capabilities of the older machine were superior, utilization of the machine's full capacity was not anticipated within the near future.

Table 1 is a comparison of the IBM-4331 and IBM System 36 computers. The System 36 computer performs all of the functions of the older machine. Many of the report and billing functions of the older machine are now performed during off hours by the System 36. Loss of speed and computing power is only marginally noticeable.

As can be seen from Table 1, salvage value of the old machine paid for the newer machine and provided a cost savings of close to $50,000 per year.

The System 36 machine is utilized more productively than the older machine. Very little has been lost due to system downsizing.

It is anticipated that the new machine will provide enough additional capacity to meet the needs of the Division for the next ten years. With the speed of technological advances, it is likely that upgrades and/or replacement will become cost effective during this planning period.

Table 1.
Specifications Comparison:
Mainframe vs Mini.

| MAINFRAME | MINI |
|---|---|
| IBM 4331 Model 2 | IBM Sys 36 Model 5360 |
| RAM: 4 Megabytes | RAM: 1 Megabyte |
| Diskspace:  700 MB | Diskspace:  400 MB |
| Operating System Requirements: | Operating System Requirements: |
| 1. CISC | 1.  SSP (includes CISC-USAM capability) |
| 2.  USAM | |
| Operating Language: | Operating Language: |
| Cobal | Cobal |
| Occupied Space .8MB | Occupied Space .1MB |
| Standard Language:  NA | Standard Language:  Basic |
| erminal Support: | Terminal Support: |
| Data Processing 3 | Data Processing 3 |
| Design          0 | Design          1 |
| O&M             1 | O&M             3 |
| Dist. 2 TP      0 | Dist. 2 TP      3 |
| Dist 3 TP       0 | Dist. 3 TP      1 |
| Accounting      2 | Accounting      2 |
| Billing         2 | Billing         2 |
| Office Mgmt.    0 | Office Mgmt.    1 |
| Other           1 | Other           1 |
|      Total      9 |      Total     17 |
| Stand Alone Machines - 0 | Stand Alone Machines - 13 |
| Printers - 1 | Printers - 19 |
| Annual Software Leasing Cost: | Annual Software Leasing Cost: |
| $16,300.00 (Perpetual) | $10,000 (one time only) |
| Annual Maintenance Cost: | Annual Maintenance Cost: |
| $18,240.00 | $7,000.00 |
| Annual Consultant Fees: | Annual Consultant Fees: |
| Updates: | Updates: |
| $20,000.00 | -0- |
| Word Processing Option | Word Processing Option: |
| Not Included | "Display Write 36" Used by O&M, Design, Office Mgmt., Accounting |
| Space Requriements: | Space Requirements: |
| 800 sq. ft. | 200 sq. ft. |
| Power Requirements: | Power Requirements: |
| 240 V 3 phs. | 240 V 1 phs. |
| Climate Control | Climate Control |

(Air Conditioning)
Continuous Operation

(Air Conditioning)
Intermittent operation

Salvage Value of Machine:
$100,000

Purchase Cost of Machine:
With 5 display stations and
3 printers
$75,000.00

The biggest advantage, beyond the cost savings, has been the addition of eight work stations. All Division departments now have access to the System 36 computer. Through emulation software (File Support Utility and Ideacomm), work stations are now provided with personal computers. These may be used for System 36 access or in the personal computer mode. The combination, mini-microcomputer network has greatly enhanced user ability to design and produce custom reports. Most of the Division's microcomputer memories (RAM) are enhanced to 640K and are provided with a 10 MB hard disk for data storage (IBM-XT).

These machines provide a powerful business computer capable of using all the latest software. Each operator has immediate access to the use of spreadsheets, databases, word processors and communication utilities. The machines also provide considerable data storage and are extremely flexible in producing desired reports.

## APPLICATION

### Automated Meter Read/Billing System

Automated meter reading begins by building a database of desired customer information provided to the meter reading device. An IBM-XT in combination with "Dataease" software is used to create the customer records.

Table 2 illustrates the entry form that is used to create the database. The meter identification number ties the individual record to the meter, which is coded with the identical number.

Access to the database is by menu. Updates to the database are made by supervisory and clerical personnel having no formal training in computer operation.

Meter reading is accomplished by loading route information into a 128K solid state interrogator (SSI). The SSI automatically enters a meter reading keyed to the meter identification number from the meter types listed in Table 3.

Other meters must be read manually and the data manually entered into the SSI unit. Downloading of the SSI meter data is accomplished by coupling the unit into a charging stand connected to an IBM-XT.

Several reports may be generated to screen the collected data using the "RMMS" system. A portion of these reports presently used by the Division is found in Table 4.

The meter reader has an option to use up to 999 predescribed note codes. Presently the Department uses 41 for field identification of problems. These are found in Table 5.

Use of the RMMS system has resulted in increased efficiency. Typically route sizes have been doubled due to the increased speed at which meters may be read with

## Table No. 2. RMMS Master Meter Reading Accounts

Community 17 Account 0000659560
Address:      6072 Flushing Road
Meter Number..................1
Meter Identification.....065956
Meter Type (B.E.M)............M

Meter Route/Sequence...037/0010
RMMS Meter Read Field:
  Manual read digits..........6
  (R)ight, (L)eft Justify.....L

B = Rockwell - ECR
E = Neptune 3 - Wire
E = Neptune - Wire
M = Manual Reading

# of low order zeros........0
Warning Message...........DOG
Reserved....................0000-00-0000-0000

Comment Field 1: 5/8 NO ARB
Comment Field 2:
Comment Field 3:

## Table 3. Meters That May Be Automatically Read Solid State Interrogator (SSI).

| MANUFACTURER | DESCRIPTION |
|---|---|
| Rockwell | TTR |
| Rockwell | ECR |
| Neptune | Tri Seal |
| Neptune | Trident 8 |
| Neptune | Trident 10 |

## Table 4. "RMMS" Reports Automation Meter Reading.

| NO | REPORT | DESCRIPTION | DATA GATHERED |
|---|---|---|---|
| 1. | Master | Provides a listing of all data in chronological order | Meter Reader meter Code Customer Acct. No. |
| 2. | Statistical Summary | Lists statistical data | No. of reads taken Avg. Time Between Reads Total Read Time |
| 3. | Route Note Report | Lists Note Codes | Note Code Account Number Address |
| 4. | Miscellaneous | Non route Reads Missed Reads Manual Reads | Account Number Address Meter Code |

## Table 5.  Note Codes.

| | |
|---|---|
| 1. Seal wire broken | 22. ARB not accessible |
| 2. Touch pad missing | 23. New account page |
| 3. Touch pad damaged | 24. No read from ARB |
| 4. Wires damaged | 25. Customer complaint |
| 5. Card left | 26. ARB pins broke |
| 6. C/B broke | 27. ARB pins bent |
| 7. C/B top gone | 28. Lower curb box |
| 8. Meter not running | 29. ARB off wall |
| 9. Meter leaking | 30. Have pool |
| 10. Water line leaking | 31. I.D. not on route |
| 11. Meter noisy | 32. Nice lawn |
| 12. Register broken | 33. I.D. incomplete |
| 13. Register damaged | 34. Building vacant |
| 14. Meter not accessible | 35. Well inoperative |
| 15. Suspect tampering | 36. No building |
| 16. Dog | 37. Move ARB |
| 17. Vicious dog | 38. No check valve |
| 18. House vacant | 39. Install ARB |
| 19. Gate locked | 40. Has ARB |
| 20. X-Connection | 41. Incomplete read |
| 21. Sump tied into sewer | |

the automated equipment.  The automated reading system has allowed the Division to retain the use of its existing billing programs on its central computer and has eliminated the need for manual meter read entry and subsequent manual data entry.

The power of the System 36 has been maintained for information storage and customer invoicing (billing), while the smaller machine (IBM-XT) is used for custom report generation.  Figure 1 is a schematic of the complete automatic billing.

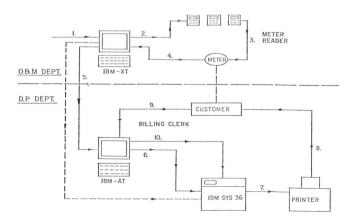

**Figure 1.**  Automated read system

1. A database is created using "Dataease."
   A transaction file is created from Dataease.
   "RMMS" imports the Transaction File & reformats it to proper protocol.
2. "RMMS" uploads the file to the solid state interrogator.
3. Data is entered into an input file in the interrogator.
4. An output file is created by the interrogator and is downloaded into the IBM-XT.
5. Meter Data (Account #'s & reads), is transferred to a "DOS" fixed length ASCII file by the IBM-XT using "RMMS."
   The "DOS" file is copied to a floppy disk and loaded into the IBM-36 terminal using an IBM-AT at another work station.
6. The "DOS" file is uploaded to the System 36 using "Ideacomm."
7. The IBM System 36 processes the data and sends the information to the printer.
8. Bills are sent to the customers.
9. The customer pays the bill.
10. The master file is updated automatically using a code reader and a precode payment card.

Errors reports (Edit Reports) produced by the System 36 are described in Table 6. Automated sytems may not be 100 percent automated due to errors which occur through accident, equipment failure or human error.

Edit reports are initially screened by the Water Department supervisor to insure timely correction of field problems.

All data entry is accomplished by the Billing Department although the department supervisor has access to all billing programs. Figure 2 is a flow schematic of the automated billing system complete with editing criterion.

Once all possible errors have been corrected and billing has been completed, it is necessary to update the customer record (master file) in preparation for the next billing period.

**Table 6.  Edit Reports.**

| Item | Report Name | Description | Information Provided |
|------|-------------|-------------|----------------------|
| 1. | Edit & Update PC to Billing | Provides Flag for Bad or Missing Data | Flag (good,bad) Account Number Input Data Last 4 consumptions Avg. last 4 consp. Existing consumption |
| 2. | Pre-billing Meter Investigation | | Flag Account Number Input data Last 4 consumptions Average consumptions Existing consumption |

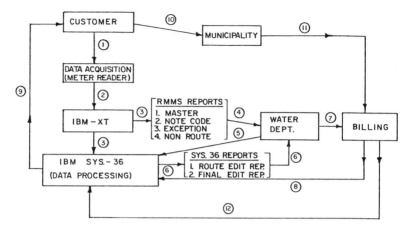

**Figure 2.**   Automatic billing system.

1.  The meter reader reads customer meter.

2.  Data is downloaded from interrogator.

    IBM-XT produces RMMS reports. Field data is transferred to System 36. Valid data updates master report.

4.  The Water Department supervisor estimates bad or absent reads. transmits information to billing and prepares work orders for field correction of problems.

5.  Correction data is entered manually.

6.  The billing program is activated and flags absent or improper reads.  The route edit report is transmitted to the Water Department supervisor.

7.  The supervisor pencils in corrections. prepares work orders and rejustifies database problems.

8.  The billing department enters the corrections and transmits the final edit report to the Water Department supervisor.  The procedure is repeated when all routes in the billing cycle are in.  A final edit report is prepared and corrections made.

9.  Individual bills are sent to the customers.

10. The customer pays the bill at the municipality's office.

11. The municipality transmits a "payment card" to the Billing Department.

12. The Billing Department automatically updates master file using a code reader.

Figure 3 illustrates the flow pattern for collection of revenue from the customer (individual bill) and from the municipality (bulk bill total) by the county agency.

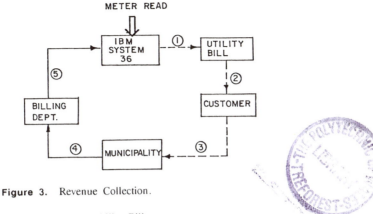

**Figure 3.** Revenue Collection.

1. IBM-36 Generates Utility Bill.
2. Utility Bill is sent to customer.
3. Customer pays municipality.
4. Municipality sends payment card to billing dept.
5. Billing dept. scans card and data is entered, automatically updates master file.

Sewer billing is accomplished in the same manner as water billing for metered sewer customers. The amount due for sewer is computed according to specific algorithms residing in the System 36 according to local ordinance.

Unmetered customers are billed according to flat rate tables established by local ordinances.

## Master Meter Efficiency Reports

The microcomputer is used to prepare monthly master meter reports which gauge the accuracy of each meter dial compared to previous months and similar periods of previous years. Monthly and mid-monthly meter readings are entered directly from the field by a billing clerk onto an IBM-XT.

Utilizing "Symphony" software, reads are transferred from the field read form directly onto an equivalent computer generated entry form. Entry onto the computer form automatically creates a database in the "Symphony" spreadsheet environment. The data is then sent to other locations in the spreadsheet using the "macro language" capability. Calculations and desired reports are generated automatically. The use of these "Symphony" capabilities simplifies meter read entries and eliminates the need to remember complicated routines. Use of this software allows the operator great flexibility in number crunching and report generation, including graphic representation. Metering problems may be quickly identified and repairs made expeditiously.

Data from previous read periods is stored on a 10 MB hard disk, facilitating file combination operations.

Use of the software's macro language capabilities greatly automates the number manipulation and report generation routines and frees the operator from remembering complicated routines. Symphony updates now allow construction of macro libraries so that once a routine is created it may be used in other spreadsheets without the necessity of retyping it.

Use of the software has allowed the division to keep its wholesale and retail water losses at about 5 percent.

Figure 4 illustrates the use of the "Symphony" software for number crunching and report generation.

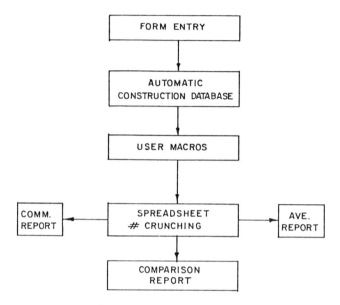

**Figure 4.**   Use of "Symphony" software for number crunching and report generation.

## OTHER USES OF THE MINI-MICRO COMPUTER COMBINATION

### General

Efficiencies have been realized in other departments as an offshoot of the computerized billing system. Several examples of the use of the computer in billing related areas are presented.

### Valve Witnessing

The Water Department uses the System 36 to keep an inventory of all valve witness information. Valve data is entered directly from valve witness field reports by the Engineering Department. Information may be updated at any time by the Water Department from a different building using a IBM-XT. Updated reports and maps are provided to each Water Department vehicle.

Customer shutoff valve locations are entered by a clerk typist onto the IBM-XT. This program is menu driven and integrated into the "RMMS" system using "Dataease."

Dataease is user friendly software which creates a database through form entry for easy retrieval. The software allows for record searches using various function keys which scan through the records using portions of data contained in any field. Records may be located using a portion of an address or a name. Scrolling backward and forward through the records is permitted.

## Inventory and Inventory Control

Water and Sewer Deparment inventory is maintained by a system of descriptive tags. Each item or group of items is described with a tag. As the item is used, the tag is removed and returned to the department supervisor, who is responsible for updating the department's inventory.

The record contains a listing of vendor information, price of item, date of purchase, number of items purchased, safety stock and reorder quantity for maintaining stock.

Another application has been the incorporation of communications software. Major parts inventory has been greatly reduced by using stock as department inventory.

This has been accomplished by the use of a "Hayes" modem and "Smarteon II" communication software which allows direct access to over 100,000 parts which are stocked in the Flint and Detroit areas. Orders may be placed automatically and delivery is guaranteed within 24 hours after placing an order. Cross references to competing manufacturer's stock are included.

Contract reports are furnished yearly for the purpose of identifying purchases and equipment descriptions.

Use of the system results in savings in storage space and parts inventory costs to the Division.

## Preventive Maintenance

Preventative maintenance tasks are performed on facility equipment by computer generated maintenance orders. A system of several hundred maintenance orders are matched to each piece of equipment. Orders are printed out by desired frequency. The department supervisor may update or change the description of the order or its frequency. Additional equipment may be added at any time. The maintenance program resides on the System 36.

## Budget Control

Budget control is maintained by a coding system which identifies each purchase by department and fund description.

All vendor invoices are coded by department supervisors. The Accounting Department enters the coding onto the main computer. Program software separates allocated and direct costs. Purchases are compared against budget allocations on a monthly basis. Reports are generated through accounting by the System 36.

## Word Processing

The Design Department makes extensive use of the System 36's "Displaywrite" software for specification preparation.

The secretarial staff uses the software for preparation of documents requiring extensive rewrites and changes.

## CONCLUSION

Medium sized utility operations can be automated by a combination of minicomputers, microcomputers and solid state interrogators at a significant savings in costs over mainframe computers and manually keypunched operations. Tangential benefits due to increased numbers of machines at work stations will result in improved efficiencies in other utility departments outside the billing area.

This chapter has provided several examples of the use of these machines, but is by no means all inclusive. Improvements in all departments will continue as the operators become more experienced and adept in using their machines.

## ACKNOWLEDGMENTS

Rockwell International - Municipal & Utility Division, 400 N. Lexington Avenue, Pittsburg, PA 15208 - and the Detroit Ball Bearing Company - Flint Office, 404 Kelso, Flint, MI 48503 - for the use of the Automated Ordering System

## SOFTWARE MENTIONED

| NO.      COMPANY | USE |
|---|---|
| 1. "Symphony" | Database |
| Lotus Development Corp. | Spreadsheet |
| 161 First Street | Report Generation |
| Cambridge, Ma. 02142 | |
| | |
| 2. "Dataease" | Database |
| Software Solutions, Inc. | Report Generation |
| 305 BIC Drive | |
| Milford Ct. 06460 | |
| | |
| 3. "Idecomm" | Emulation |
| Ideassociates, Inc. | |
| 35 Punham Road | |
| Billerica, Ma. 01821 | |
| | |
| 4. "Smartcomm II" | Communications |
| Hayes Micro | |
| 5923 Peachtree And Blvd. | |
| Norcross, Georgia 30092 | |

5. "RMMS"                                         Automatic Meter Reading
   Rockwell international                          Report Generation
   Municipal & Utility Division
   400 Lexington Avenue
   Pittsburgh, Pa. 15208

6. "Displaywrite 36"                              Word Processing
   IBM Corporation
   Dept. 52 A
   Neighborhood Road
   Kingston, N.Y. 12401

7. "System Support Program"                       System 36 -
   IBM Corp.                                      Operating Language
   Dept. 52 Q
   Neighborhood Road
   Kingston, N.Y.

8. "File Support Utility"                         PC to Sys 36
   IBM Corp.                                      File Transfer
   Dept. 52 Q
   Neighborhood Road
   Kingston, N.Y. 12401

# CHAPTER 8

## UTILITY RATE STUDIES - DEVELOPMENT OF USER CHARGE SYSTEMS

David J. Vaclavik, P.E.
Camp, Dresser & McKee
Detroit, Michigan

## INTRODUCTION

The development of utility rates has been greatly aided by the progress that has been made in the computer industry. Calculations formerly made by hand with the use of a calculator or adding machine are now more effectively made by powerful computers capable of performing multiple operations. Many of the microcomputers today actually cost substantially less than some of the early calculators. Although rate methodologies have been programmed and installed on mainframe and minicomputers for many years, it was the development of the spreadsheet model Visicalc and the many other models since then, that made microcomputers a practical choice for the development of rate models. There are now literally dozens of spreadsheet models available for virtually every type and size of microcomputer.

Anyone familiar with rate studies is aware of the large number of calculations that are performed in achieving the final rates. Most of these calculations involve tabular data for which spreadsheet programs were specifically developed.

Although more expensive, spreadsheets also have been developed for larger computers. Many of the office automation systems being established include a spreadsheet program that is fully adequate for rate development. These systems also provide the added advantage of being able to integrate with text generated on the word processing system and to access information stored in database files. These systems are becoming more available as the price of minicomputers continues to decline.

## OVERVIEW

This chapter will include a discussion of the use of computer systems to design and analyze potential utility rate methodologies. The primary focus will be on the use of microcomputers and spreadsheet applications to develop rates. Some attention will also be given to larger systems using either other spreadsheets or specific programs written with a programming language such as Fortran or Pascal. The advantages and disadvantages of these approaches regarding ease of maintenance, development time, and functionality are discussed. Primary advantages of a microcomputer-based spreadsheet model are low cost, availability, ease of development, the ability to change variables and assumptions easily, and the comfort and familiarity that many users have developed with these systems. The interactive style provided by a spreadsheet makes it convenient to quickly put together alternative solutions and test ranges of variables. Rate analysis includes a number of distinct tasks:

- determination of revenue requirements
- analysis of usage/discharge characteristics
- development of user classes
- assignment of costs to classes
- calculation of rates

Each of these tasks can be assisted by the use of a computer system. Development of user classes and rate blocks and the identification of total flow volumes can often be done best by analyzing the data accumulated through the computer billing system. The cost information required to develop rates is essentially budget information. Therefore, properly developed, a spreadsheet rate model will be of great value not only in determining rates, but also in preparing the annual budget.

## PREREQUISITES

Several issues need to be resolved in making the decision to develop a computer-based rate model. These can be separated into two basic categories: "Why should I" and "Can I." In general, a computer model designed for rate analysis becomes more appropriate as the complexity of the rate structure increases. If the user has any familiarity with computers, simple rate structures can be established on spreadsheets with minimal efforts.

Once established, a computer-based rate model makes the process of updating rates simpler and allows the user to determine, at any time, the results of changes in assumptions on projected revenues. The "Can I" part of the question revolves around both the individual's particular abilities and the second basic issue, computer availability. Assuming that the user has an understanding of rate-making practices and some familiarity with computers, there is no reason why, given access to appropriate facilities, the user can't develop a computer model on his or her own.

This second basic issue, computer availability, requires an understanding of the resources necessary to develop a rate model. The two separate components required for rate development are:

- computer hardware
- computer software

These two components are discussed in the following paragraphs.

## HARDWARE REQUIREMENTS

Although the distinctions are beginning to blur, there are three generally accepted classes of computers. These are:

- microcomputers

- minicomputers
- mainframes

In general, all three classes of computers are suitable for the development of rate models. For the purpose of this chapter minicomputers and mainframes will be considered as one category and microcomputers as another category.

## Mini/Mainframe Computers

Mini/mainframe computers are typically multiple user and operate at speeds greater than that provided by microcomputers. At a municipal level, this class of computer is frequently associated with accounting functions or office automation systems. The suitability of these systems for use in developing utility rates is almost completely software dependent. Virtually all machines in this class have sufficient power and capacity for rate modeling.

## Microcomputers

A great many different brands of microcomputers are offered by various manufacturers. Some of the more commonly used include the IBM PC family, the Apple IIe and Macintosh, and the Dec Rainbow. Most of these systems are suitable, with proper software, for developing rate methodologies. The limiting factor on microcomputers in determining their usefulness in modeling is usually related to available memory. Exact memory requirements to develop a working rate model are completely dependent on the complexity of the model and the requirements of the development software. Simple systems may be developed on machines with 64 K of memory or less, whereas more complex systems may require memory in excess of 512K.

## SOFTWARE REQUIREMENTS

Generally, all three classes of computers are capable of supporting the calculations required to perform rate analyses. The availability of appropriate software is usually a more important concern to the prospective rate modeler. Two software approaches are discussed in this chapter. The first is the use of spreadsheet software such as Lotus 1-2-3, Supercalc, and many others. The second approach is using a programming language to create a model specifically for utility rate analysis.

## Spreadsheet Models

Spreadsheet programs, whether established on a microcomputer or on a larger system, are ideally suited for the manipulation and analysis of the large quantities of data associated with rate development. Next to word processing, spreadsheets are the most commonly used software. A large number of spreadsheet programs have been developed to run on microcomputer sytems. Most are adequate for rate development and the choice of which to use is a matter of personal taste.    The prices for these systems begin at less than $100.

Spreadsheet programs also are available for larger systems.    The selection is somewhat more limited and costs are generally higher; however, these programs operate very similarly to those designed for microcomputers.    As more communities begin to install office automation systems, access to these spreadsheets is becoming more common.    Many office automation systems are designed so that spreadsheet functions can be linked to the word processing and database capabilities also provided. Although some ability to do this exists with certain microcomputer packages, it is an area where larger systems may provide greater utility.

One of the primary advantages of spreadsheet software is its ability to be used as a development tool.    Once data have been entered into the system, spreadsheet programs provide interactive access to the data nad allow data manipulation to be carried out with relatively few steps.    Typical rate information such as cost data, customer descriptions, and volumes of consumption or discharge can be shown best in tabular form.    Once the base information has been entered, alternative projections and calculation methodologies can be carried out with a minimum of effort.    Once appropriate procedures are determined, the spreadsheet can be cleaned up and protected for use on future rate calculations.

## Programming Languages

Before spreadsheet software was available, developing a computer model for utility rates required writing a program specifically designed for that purpose.    Because of the effort required to do this, rate models were traditionally developed manually and then converted to programs for computer use.    Although this is fine for updating rates at future dates, it doesn't provide a great deal of help during development. For most applications, and particularly those intended for microcomputer systems, an approach built around a spreadsheet is more appropriate and offers greater flexibility.

Developing a computer program based on a specific methodology becomes practical, and sometimes necessary, under certain circumstances.    Generally, this occurs when the rate methodology:

- becomes extremely complex
- needs to manipulate large quantities of data
- must interface with other programs or applications

## ASSOCIATED APPLICATIONS

To be successful, a computer-based utility rate model must somehow access the support information necessary to produce a rate schedule.    This information will include revenue requirements, consumption/discharge volumes, and customer characteristics. In most cases, this information is entered directly by typing the data into the system at the time rates are being calculated.    In developing a computer-based rate model, it is important to evaluate the sources of the information that will be required to determine the rates.

Information concerning revenue requirements is generally available from budget worksheets.    Customer information such as meter size, class, and consumption/ discharge volumes is commonly obtainable from the billing system.    Other information may come from plant operation reports.    All of these information sources are increasingly being managed with computer systems.    The key is in making sure that each source supplies that information needed by the utility rate model.    An ambitious system may access computer files prepared by each of these programs directly.    Generally, the required information would be generated in report form and then manually entered into the rate model.

## Budgeting

To develop rates that will generate the revenues necessary to keep the utility system in good fiscal condition, it is first necessary to determine what the revenue requirements for the system are.    This process generally occurs annually at budget time.    The level of detail produced by the budget process determines, to a certain extent, the degree of complexity and equity that can be built into the rate model. The more information provided by the budget, the easier it becomes to develop an equitable system of rates.

Both budgeting and rate modeling can benefit greatly from the use of a spreadsheet program.    In fact, the budget information worked out this way can be the first step of the rate modeling process.    The key is in providing sufficient details to allow assignment of costs appropriately for rate development.    The need for detail is completely dependent on the rate methodology; however, typical cost identification categories may include:

- meter reading/billing costs
- capital improvements
- distribution/collection system costs
- treatment costs associated with high strength (surchargeable) wastewater discharges
- costs associated with peak flows
- fire protection
- administration

## Capital Improvement Planning

Capital improvement planning is the process whereby a utility determines what construction activities will be carried out and when.    Except where funding is made through special assessment or general levy, capital improvement costs directly affect the rate structure.    In many cases, capital costs become one of the major components of a rate system.

One of the factors in deciding on and particularly in scheduling capital improvements is their effect on the rate system.    Small improvements to be paid out of revenues generated during the year can be input into the rate model easily.    The problem comes in determining the effect of capital improvements that require long-term financing.    In this instance, capital improvement costs need to be expressed as a string of annual costs rather than as a total.    Conversion from project cost to an annual bond payment requires making a series of assumptions with respect to financing costs, interest rates, term, and a host of other factors. This calculation can either be done independently or be provided for in the rate model itself. Regardless, capital financing is a factor that needs careful consideration and one that will have significant impact on any utility rate model.

## Billing Systems

One of the most important sources of information for rate development can be the utility's computer billing system. This system normally maintains all of the customer data, such as:

- number of accounts
- meter sizes
- customer classes
- consumption/discharge volumes

Unfortunately, many billing systems are designed only for managing billing information in terms of accounts due, accounts paid, and total revenue. Getting the desired information out of some billing systems can be challenging. With new database systems becoming popular, this problem should decrease; however, it often will be necessary to either write a program to retrieve and analyze the desired billing data or to modify the billing program itself so that it provides the necessary information.

## Plant Operation Reports

Plant operation reports provide several pieces of information of benefit to the rate modeling process. These reports provide the required data on volumes of water or wastewater produced or treated of as well as information on each of the treatment processes. Total quantities surchargeable pollutants are identified and an estimate of inflow/infiltration or line loss amounts can be made by comparing with metered customer totals.

## UTILITY RATE MODELING GUIDELINES

A number of issues related to computer modeling must be considered when developing a computer-based utility rate model. Any of these can seriously affect the usefulness of the program. Some of these issues are:

- documentation
- testing
- user reliance on program

## Documentation

There are two parts to every utility rate study.   One is the series of calculations leading to a new rate structure.   The other part, which is equally important, is the written discussion and explanation of how the rates were developed, what assumptions were made, and the logic behind each decision or formula included in the rate methodology.

Without adequate documentation, a rate model may end up causing more difficulty than it is worth.   When working with a computer model, good documentation is essential and is actually required in more detail than it would be with a manual or "hand" calculated system.   To be complete, the documentation should include:

- the rate methodology
- operation instructions for use of the system
- program structure

## Rate Methodology

The documentation provided should be sufficient so that rates could be updated independently of the computer model if so desired.   All assumptions, calculations, and tables associated with the rate procedure should be included and discussed in the documentation.   Someone reading through the documentation should be able to understand how the rates were developed and the logic behind the methodology chosen. This minimum level of understanding is critical if the rate model is to be updated and used later, because assumptions made during the initial development of the rate methodology may not be appropriate at future dates.   The only way to determine this and make appropriate modifications is through careful documentation.

## Operation Instructions

No matter how simple the operation of the model seems to the developer, complete directions need to be written documenting the steps necessary to run the Utility Rate Model.   These directions begin at how to load the program and continue through every step required to complete the rate calculation process.   Any program "help" features that have been built in need to be identified, as do any constraints set on variables.   One of the major problems with interactive spreadsheet models is the ability that exists to accidentally erase a key formula or piece of information that will either distort the rates or cause the model to malfunction altogether.   Some spreadsheet models can protect against this; many do not.

One of the areas of documentation that requires particular attention is the recognition that people make mistakes.   Frequently, user documentation tells you how to do something correctly, but is vague on how to recover once an error is made.   In certain cases, it will be possible to go back to the place the error was made and correct it there.   If a formula has been erased or altered, it may be easier to reload the program and begin over again.   One of the problems with any series of calculations is in recognizing that a mistake has occurred.   This is particularly true when the mistake is minor.   With a spreadsheet approach to rate modeling, it is possible to check column and row totals fairly easily.    These totals can then be compared against source documents as a quick check.

Depending on the capabilities of the spreadsheet model or program being used, it may be possible to have the rate model check for certain errors.    One of the more common error checking techniques is to set allowable ranges for data that will be entered.    When this is done, any attempt to enter data outside the acceptable range is rejected.    This capability can be built in with certain spreadsheet packages and can always be written into a model developed with a programming language.

## Program Documentation

At some point in a computer model's life it becomes necessary to modify the code. These modifications may be as simple as adding a new line item in the budget information, or they may entail major changes in the formulae and assumptions incorporated into the rate model.    The point is, there will be a need to be able to modify the program.    With spreadsheet models, revisions usually can be made quite readily.    However, it is always a good idea to maintain documentation showing each set of equations and its relationship to other equations.    Some spreadsheets allow a user to print all formulae, resulting in a listing of the equations and identification of what and where the inputs to these equations exist in the spreadsheet.

As any programmer can tell you, the issue of program documentation is an extremely serious one if the model has been written with one of the many programming languages. First, the program can be revised only by someone familiar with the programming language used.    Even if the program is written in a familiar language, an undocumented program may be very difficult to modify.    It is extremely important that each subroutine or calculation be documented in the program with sufficient detail to allow another programmer to follow the code and make necessary revisions.

## Testing

As with any computer model, a utility rate program should be thoroughly tested before it is used.    In addition to checking that specific data result in correct rates, each of the features built into the system should be completely tested. Acceptable data ranges that have been built in should be checked to see that data outside the range results in the desired response.    "Help screens." if and where provided, should be reviewed to make sure they can be properly accessed.    Finally, it is likely that certain keystroke combinations or data entries will cause the system great difficulty.    Where the program can be modified to avoid these, it should be. Where such problems are likely to continue, the limitations of the modeling system should be identified and discussed in the operations manual.

## User Reliance

One of the most dangerous aspects of a computer model is that eventually someone considers the output to be gospel.    The justification of something on the basis of computer analysis is only as valid as the computer model being used. Particularly with rate models, it is important to realize how the system is designed   and that assumptions made during the rate development may need updating.   As long as a computer model is left alone, it will continue to turn out rate structures using the same cost assignments, formulae, and logic that were designed into the system.   Once the logic changes, the model must be revised or the rates developed will no longer reflect actual conditions or be equitable to the system's customers.

## EXAMPLE RATE MODEL

The example rate model shown and discussed in this section is greatly simplified and is intended only to demonstrate some of the principles discussed previously.   It does not necessarily represent accepted rate-making policy, nor is it intended to. The model shown was developed using Lotus 1-2-3, a product of the Lotus Development Corporation.   As with most spreadsheet rate models, the information is presented in table format.

Table 1 gives the user the opportunity to make certain assumptions with respect to the information to be used calculating the rates.   Each of these assumptions affects the following tables and calculations, and therefore the final rate.   This model is set up with 1984 as a base year and calculates rates for 1985.   Table 1 allows input as to inflation rate, customer growth, and changes in the billable flow.   The inflation rate can be set generally for all cost categories.   It also provides the opportunity for setting specific rates for personnel and utility costs.   In this example, the general rate was set at 5% and the rates for personnel and utilities were set at 6% and 10%, respectively.   Customer growth is entered in real numbers representing projected new customer accounts.   Percent change in billable flow allows the user to establish specific increases or decreases in flow estimates for each of the three user classes.

Table 2 shows the increased revenue requirements determined by applying the selected inflation rates to 1984 costs.   If the inflation rates had been left blank for personnel and utilities in Table 1, the program would have automatically used the general inflation rate.   In this model, an inflation rate was applied to debt service costs.   Generally these costs are well established and would be entered directly. Table 3 is designed to make billable flow projections based on the assumptions in Table 1.   The 1985 number of customers reflects the 1984 number plus the changes specified in Table 1.   The model has been set so that the 1985 billable flow is calculated based on the percent increase specified in Table 1.   If no rate is specified, as with the commercial and industrial accounts, the billable flows are modified in proportion to the change in the number of customers.

Table 4 is intended to calculate meter equivalents to be used in establishing meter charges.   The number of equivalent meters is calculated by multiplying the meter ratio for a given meter size by the number of meters of that size.   The program has been designed to verify that the number of meters shown in Table 4 checks with the total number of customers from Table 3.   If it does not check, as in Table 4A, an error message is printed.

Table 5 shows the calculation of an equivalent meter charge based on administrative costs plus 50% of debt service.   These costs are totaled and then divided by the number of equivalent meters from Table 4 to get a charge per equivalent meter.

Table 1.  Assumptions.

| | |
|---|---|
| Inflation Rate | |
| General | 5% |
| Personnel | 6% |
| Utilities | 10% |
| Customer Growth | |
| Residential | 75 |
| Commercial | 11 |
| Industrial | 3 |
| Percent Change Billable Flow | |
| Residential | 10% |
| Commercial | |
| Industrial | |

Table 2.  Revenue Requirements.

| | 1984 | 1985 |
|---|---|---|
| Personnel | $195,000 | $206,700 |
| Utilities | $45,000 | $49,500 |
| Chemicals | $27,000 | $28,350 |
| Administrative | $48,000 | $50,400 |
| Debt Service | $85,000 | $89,250 |
| Miscellaneous | $23,000 | $24,150 |
| TOTAL | $423,000 | $448,350 |

Table 3.  Billable Flow Calculations.

| | 1984 | | 1985 | |
|---|---|---|---|---|
| | No. of Customers | Billable Flow | No. of Customers | Billable Flow |
| Residential | 3,200 | 192,000 | 3,275 | 211,200 |
| Commercial | 300 | 22,500 | 311 | 23,325 |
| Industrial | 50 | 12,500 | 53 | 13,250 |
| TOTAL | 3,550 | 227,000 | 3,639 | 247,775 |

Table 4.    Equivalent Meter Calculations

| Meter Size | Meter Ratio | Number of Meters | Number of Equivalent Meters |
|---|---|---|---|
| 5/8 | 1 | 3,425 | 3,425 |
| 3/4 | 1.5 | 114 | 171 |
| 1 | 2.25 | 80 | 180 |
| 1-1/2 | 4 | 15 | 60 |
| 2 | 6 | 5 | 30 |
| | | 3,639 | 3,866 |

Table 4A.    Equivalent Meter Calculations.

| Meter Size | Meter Ratio | Number of Meters | Number Equivalent Meters |
|---|---|---|---|
| 5/8 | 1 | 3,425 | 3,425 |
| 3/4 | 1.5 | 114 | 171 |
| 1 | 2.25 | 80 | 180 |
| 1-1/2 | 4 | 15 | 60 |
| 2 | 6 | 6 | 36 |
| | | 3,640 | 3,872 |

ERROR METER COUNT DIFFERS FROM CUSTOMER COUNT

Table 5.    Calculation of Meter Charge.

| | |
|---|---|
| Administrative Costs | $50,400 |
| 50% of Debt Service | $44,625 |
| TOTAL | $95,025 |
| Number Equivalent Meters | 3,866 |
| Annual Charge per Equivalent Meter | $24.58 |
| Quarterly Charge per Equivalent Meter | $6.14 |

Table 6.  Quarterly Meter Charges.

| Meter<br>Size | Meter<br>Ratio | Quarterly<br>Meter Charge |
|---|---|---|
| 5/8 | 1 | $6.14 |
| 3/4 | 1.5 | $9.22 |
| 1 | 2.25 | $13.83 |
| 1-1/2 | 4 | $24.58 |
| 2 | 6 | $36.87 |

Table 7.  Calculation of Commodity Charge.

|  | 1985 |
|---|---|
| Personnel | $206,700 |
| Utilities | $49,500 |
| Chemicals | $28,350 |
| 50% Debt Service | $44,625 |
| Miscellaneous | $24,150 |
| TOTAL | $353,325 |
| Billable Flow | 247,775 |
| Rate Per 1,000 Gallons | $1.43 |

Table 6 takes the equivalent meter charge and multiplies it by the meter ratio for each meter size to obtain the quarterly meter charges.

Table 7 contains the calculation of the commodity charge.  This example shows the calculation of a uniform commodity rate, determined by dividing the costs not assigned to the meter charge by the total billable flow.

## SUMMARY

Computer rate modeling provides a number of significant benefits including ease of development, the ability to make revisions or modifications quickly, and the ability to test the effects of a large number of different assumptions on the end rates.  A computer rate model can be developed quite easily by someone with an understanding of rate design and some familiarity with a spreadsheet model.  Computers, however, cannot overcome or correct a poorly designed rate structure and will be beneficial only as long as they are used and updated with care.

# Chapter 9

# Water Network Analyses

Paul Shay, P.E.
Wade, Trim & Associates
Taylor, Michigan

## INTRODUCTION

Analyses of potable water distribution systems are routinely accomplished by the use of computer modeling. The computer program solves a set of hydraulic equations that will predict pressures and flows throughout the system under consideration. While very powerful computer programs developed for this purpose have become increasingly available in recent years, it remains the responsibility of water system managers and engineers to properly define input parameters, calibrate the system model, and interpret model results.

The principle items to be addressed in any system analysis are summarized as follows:

- data collection
- formulation, calibration, and verification of system network model
- design criteria determination
- computer analyses
- master water plan development.

## DATA COLLECTION

Paramount in the analysis of a distribution system is intimate knowledge of that system's composition. Too much emphasis, then, cannot be placed on the importance of thorough data collection. Typically, a municipality or water system manager will be able to furnish data such as system maps, as-built construction drawings, elevation contours, manufacturer's pump curves, storage reservoir data, water supply, and sales records. The engineer, however, is required to investigate further to secure that data required to perform an accurate system analysis.

In building a system's network model, one begins by reducing the system map into a series of lines or pipes, connecting junctions where two or more pipes are connected.

For large systems, all pipes 12 inches in diameter, or larger, are normally included. Smaller, loop connecting 6- and 8-inch mains, capable of transmitting significant quantities of water, are sometimes included. The number of pipes to be included in a network model is a function of the degree of accuracy desired and the capacities of the computer hardware and software.

Pipe information required for the system analysis includes pipe diameter, length, and Hazen-Williams C-factors. Pipe length and diameter can be obtained directly from the system map or as-built construction drawings. C-factors are typically selected from accepted engineering literature based on diameter, material, and age of pipe. The C-factors are then adjusted as appropriate to calibrate the network based on the field testing results. C-factors also can be selected based on C-factor tests. Hydraulic grades at opposite ends of a known length of pipe can be measured. If the flow in that pipe can be measured, the C-factor can be calculated using the Hazen-Williams formula as follows:

$$\text{C-factor} = \left[ \frac{L Q^n 4.27}{hf\ D^m} \right]^{1/n}$$

where,

L  = length in feet
Q  = flow in cfs
hf = head loss in feet
D  = diameter in feet
n  = 1.852
m  = 4.870

C-factor tests, however, are difficult to implement in normal distribution systems. Isolated lengths of mains are difficult to establish and flows are difficult to accurately measure without severe disruption of system operations. The selection of accurate C-factors has been, and continues to be, the source of the greatest uncertainty in collecting system data. It has been found that textbook values for pipe C-factors, with modifications resulting from field test calibrations, are sufficient for normal system analyses.

Junctions or nodes define points of pipe interconnections. Additional junctions can be included to define hight points, low points, or locations of major water consumption. Data needed for each junction are the pipe connectivity, ground elevation, and system demand. The system demand for any particular junction is most easily determined from the water billing data for that junction's influence area. For large systems, customer service meters 1 inch and larger are normally identified and their consumption extracted from the billing data. The remaining consumption should be separated by use category, i.e., residential, commercial, industrial, office, or system losses.

Residential consumption can be estimated by multiplying the population contained in each junction's service area by a predetermined consumption rate. The consumption rate will vary with factors such as density, age, and amount of sprinkled lawn areas and can range from 70 to 130 gallons per capita per day. Several typical areas should be analyzed in great detail to determine the appropriate consumption rate to be used for the entire area.

Commercial, industrial, and office usage are often grouped together and a consumption rate is similarly determined. The rate can be expressed in gallons per acre per day, or in gallons per square foot of floor space per day. Rates computed by the author for medium to large municipalities range from 1,000 to 3,000 gallons per acre per day.

System losses are calculated from water treatment plant and billing records and are simply the difference between water produced and water sold. System losses of 10-15% are normally encountered. Without conducting an extensive water use audit,[1] losses can be distributed equally among the network's junctions. This will result in higher losses in areas of older, more congested development with a high junction density and lower losses in newer areas where junctions are more distant from each other.

The City of Ann Arbor, Michigan has developed a rather novel, time saving method of collecting this data. The service area was divided into a series of polygons with only one junction in the center of each polygon. Each customer meter was assigned a 3-digit code based on its location in the system, corresponding to the system junction number. As new customers are added, their location on the polygon map is determined and a junction code assigned. For any desired period of time, consumption quantities for a particular junction code are easily retrieved from the billing data stored on the city's main computer system. In this manner, consumption rates for each area are accurately determined.

Data for ground and elevated storage reservoirs are readily available. Information required for storage reservoirs include high-water level, low-water level, capacity, physical dimensions and adjacent ground elevations. Conversations with water treatment plant personnel also are useful in determining the basis for reservoir operation, including piping configurations for filling and emptying.

Performance curves for high-service pumps also need to be obtained. Usually these curves will be furnished by water treatment plant personnel. If lost, pump manufacturers can be contacted to secure the required data. It should not be assumed that pumps are operating on the pump curves supplied, even if they are certified. Pump curves should be verified by field testing, especially if they are ten years old or older. Flows can be measured by a recently calibrated flowmeter or by using a pitot tube. Suction and discharge pressures can be measured directly using standard pressure gauges. Several operation points can be observed and a comparison made with the pump curves supplied.

Location and status of all valves in the system should be documented and verified. Pressure reducing valve settings can be checked by reading accurate pressure gauges upstream and downstream of the valve.

One cannot stress enough the importance of using accurate meters and gauges in collecting the required data. Prior to beginning a water system analysis all pitot gages, static gauges, and flowmeters should be calibrated to ensure accurate results.

## MODEL CALIBRATION

Field testing is required to verify or calibrate a computer network model. Several strategic locations throughout the system are selected and static and residual pressures measured. Residual pressures are measured at a fire hydrant with adjacent hydrants flowing.[2] The intent is to stress the system sufficiently to produce a minimum of 10 psi pressure drop at the test hydrant. In addition to the pressure and flow data collected at the test location, it is imperative that data throughout the system be collected at the time of the test. The information required includes system flows, pressures, reservoir levels, status of valves, and high service pumping configuration.

Using the network model, flows and boundary hydraulic grades are input.    The field tests are then simulated on the computer.    Comparison of the computer results with field measurements indicate the accuracy of the model.    Static pressures within 3 psi and residual pressures within 5 psi of predicted results indicate that a fairly accurate model has been developed.    Results outside these ranges require model calibration.    Each system will dictate differing methods of calibration.    It is the engineer who determines which of the collected data has the greatest degree of uncertainty.    The highest degree of inaccuracy normally lies in the determination of consumption levels and/or C-factors.    Walski[3] discusses possible inaccuracies found in network modeling and suggests several calibration methods.    It has been the author's experience that slight modifications to pipe C-factors will yield satisfactory results if flows have been derived from sound billing data and boundary conditions have been accurately defined.

## DESIGN VALUES

With the network model satisfactorily developed, the next step is to define parameters with which to analyze the distribution system.    The most important parameter is the determination of the study period to be designed for and analyzed. For new systems with capacity for large expansion, design periods of 5, 10, or 20 years are common.    For larger or older systems, where service area boundaries have been defined, an ultimate design period may be appropriate.    At ultimate development, it is assumed that the study area will reach its maximum population and all vacant lands within the service area will be fully developed.    In either case, the designer will have to rely on the current Master Land Use planning document for the municipality being examined.    This document will help define the quantity and distribution of future water consumption.

Existing demands and their distribution throughout the system are determined in the model phase from examination of the billing data or other methods.    Maximum hour, minimum hour and maximum day peaking ratios can be gleaned from water treatment plant records.    Maximum day flows are routinely recorded for most water systems.    Maximum and minimum hour figures can be calculated by examining pumpage and reservoir level recording charts for several weeks either side of maximum day.    Knowing reservoir capacity, inflow or outflow rates can be estimated by converting level changes to quantities of water for a given period.    By algebraically adding all the flows from pumpage and reservoirs, the maximum hour and minimum hour consumption levels are derived.    These levels are usually expressed as a ratio by dividing them by the average day flow.    These ratios should be reviewed and approved by the Department of Public Health or other governing agency for the study area prior to use in the analyses.    Figure 1 depicts the 1984 maximum day hydrograph for the City of Ann Arbor, Michigan.

## SYSTEM ANALYSES

Analyses of the distribution system normally consist of the following computer simulations:

- average day
- maximum day

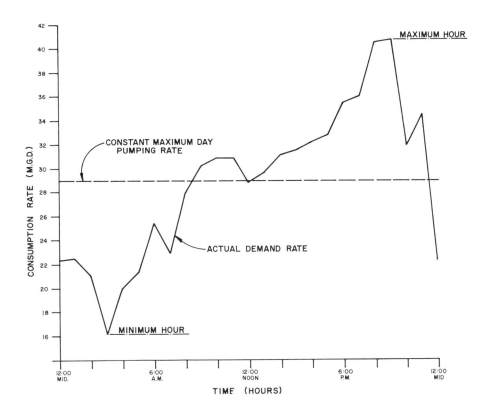

**Figure 1.** City of Ann Arbor maximum day Hydrograph (July 24, 1984).

- maximum hour
- maximum day plus fire flow
- minimum hour on maximum day (tank refill)

In Michigan, the Department of Public health required fire demand to be superimposed on "peak flow." This can be interpreted to be either maximum day or maximum hour levels. The appropriate interpretation is subject to service area characteristics such as size and composition. This selection also should be coordinated with the reviewing agency.

The existing system is analyzed for the above conditions with existing consumption levels and with consumption levels consistent with future design periods. Deficiencies are identified and improvement programs conceived.

It is desirable for the water treatment plant to have the capacity to produce and pump maximum day flows. System storage and remote pumping are expected to dampen the maximum day fluctuations and provide for fire demands and emergency reserve. The computer analyses should be examined to verify the system's ability to meet these criteria while maintaining certain pressure parameters. Allowable pressures will vary from system to system depending on customer requirements. Minimum pressures of 35, 30 and 20 psi are normally recommended for maximum day, maximum hour and fire demands, respectively. If hose streams are used directly from hydrants instead of fire pumpers, residual pressures of 50 psi and more will be required for fire fighting purposes. Maximum pressure should be limited to 100 or 110 psi to protect against residential water tank failure.

Static analyses will determine instantaneous flow and pressure parameters for that particular consumption level. A mass diagram constructed from the daily hydrograph of maximum day usage will help determine the adequacy of existing system storage. Figure 2 represents a typical mass diagram and illustrates a method of calculating required operation storage. Additonal storage will be required for fire demands and emergency reserve.

A more dynamic analysis has been recently developed, however, that allows one to more accurately determine storage and high-service pumping adequacy. The Extended Period Simulation (EPS)4 analysis is capable of dynamically changing demand levels to match the maximum day demand hydrograph, including the superimposition of fire demands. Knowing reservoir geometry, the program will project reservoir levels at various predefined intervals throughout the simulated day - normally one hour. Pressure switches located throughout the system can open and close pipes and turn pumps on and off as the system reacts to the varying consumption levels, system pressures, and reservoir elevations. In this way, one can verify a system's ability to handle fluctuations in system demands caused by maximum hour or fire demand. Maximum and minimum pressures, as well as a reservoir status summary table, are printed at the end of each interval.

It remains to ensure that the imrovement programs developed will correct the deficiencies identified in the analysis of existing and future consumption levels. If not, additional transmission, distribution, pumping, or storage improvements need to be developed.

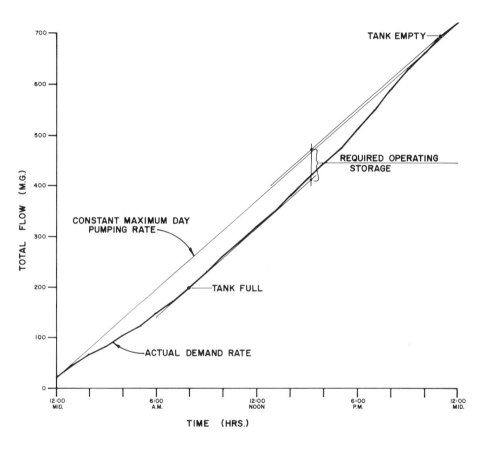

**Figure 2.** Mass diagram of maximum day.

## CONCLUSION

The purpose of the water network analysis is to provide a master plan for the operation and development of the water distribution system. The system's ability to provide adequate flows and pressures throughout the service area will have been determined.

Of equal importance will be the development of the computer model that will evaluate or predict the system's response to unexpected system demands. Pipe sizes, system storage facilities, and high-service pumping can then be sized to be consistent with the goals set forth in the master plan rather than by reactions to isolated service requests throughout the service area.

The degree to which the computer model and master plan become useful is inherently tied to the engineer's development of an accurate computer model and realistic design parameters. Great care must be exercised in collecting the data needed to define the distribution network and develop the computer model. The main responsibility lies in the ability to provide the engineer with a realistic Master Land Use Plan which defines the projected requirements of the water service area as accurately as possible. Based on the Master Water Plan, maintenance and capital improvement programs can be planned and executed with the confidence that the existing and future service area will be adequately served.

## References

1. Cori, K.A. "Auditing a Water Distribution System Points to Improvements." Water Engineering & Management, October, 1985.

2. Insurance Services Office "Fire Flow Tests." New York, 1963 (1973 Reprint).

3. Walski, T.M. "Assuring Accurate Model Calibration." JAWWA 77:12:38 (December, 1985).

4. Wood, D.J. "User's Manual - Computer Analysis of Flow in Pipe Networks Including Extended Period Simulations." Office of Continuing Education and Extension of the College of Engineering of the Univerity of Kentucky, Lexington, Kentucky, 1980.

# CHAPTER 10

# COMPUTER AIDED MAPPING AND DESIGN

Mohinder N. Kumra
Engineering and Graphic Services. Inc.
Oak Park, Michigan

## INTRODUCTION

Computers are gaining applications in the everyday work environment for ease of access of data and faster processing of information. Digital cartography is a well established science and such data is used in many high-tech applications. This chapter will introduce the rudiments of a Computer Aided Design and Drafting (CADD) system. Hardware and software requirements of a CADD system for water and wastewater applications are reviewed. Also discussed are the concepts of computer aided mapping (and drafting) and retrieval of such information for management and other design applications.

## CADD SYSTEM

Computer aided design and drafting systems provide automated mapping, design and drafting and interactive information. The interface between the graphic and non-graphic data is very essential for water and wastewater applications.

### CADD System Hardware

CADD system capacity requirements depend on the size of the water system and on the applications of the system. An average size system is comprised of the following components:

- 1 central processing unit (CPU) - a data processing device
- 2 disk drives - an active storage device
- 1 tape drive - a media transfer device
- 2 graphic terminals with input device
- 2 alphanumeric terminals - an input device
- 1 printer - an output device
- 1 plotter - an output device

Most of the software vendors configure the hardware system to their requirements.

Changes made to the standard hardware are only serviced by the software vendors. Some software packages run on standard available equipment. Figure 1 shows a typical CADD unit.

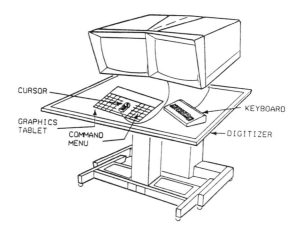

**Figure 1.** A typical computer aided design and drafting (CADD) system.

Microcomputer based systems are generally interactive drafting machines or solely Data Base Management Systems (DBMS).

## CADD System Software

The software is more important than the hardware. Software drives the system to perform certain functions. It must be able to interface between the graphic and non-graphic data. The software must perform:

- graphic communications (input/output)
- text or alphanumeric communications (input/output)
- full graphic reporting capabilities
- full tabular reporting capabilities
- data structure for facilities system modeling
- graphic and facilities data synchronization
- interface with existing operations and programs
  operations from remote locations (optional)

Figure 2 shows the interaction between graphic and non-graphic data.

Among various CADD system vendors, two vendors in particular are known for mapping. These are Intergraph and Synercom. Other CADD systems include Applicon, Calma, Computervision, and IBM. Programmability of the system to suit the department requirements should be considered very important. Use of standard hardware in the system will help increase the versatility and decrease maintenance costs.

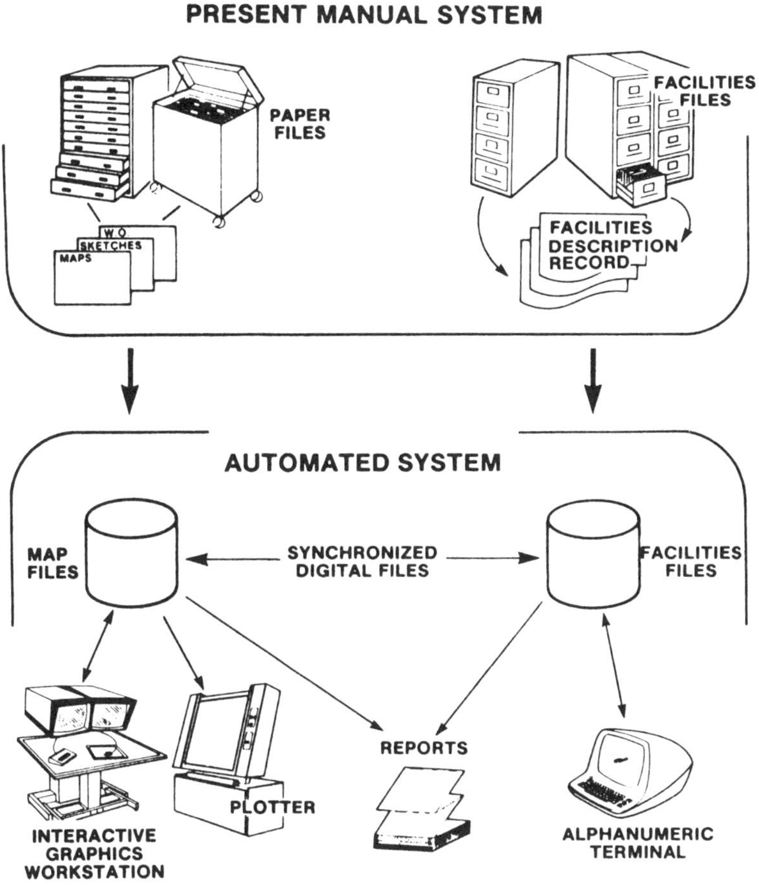

## PRESENT MANUAL SYSTEM

**Figure 2.**   Interaction between graphic and non-graphic data.

## MAPPING FOR WATER FACILITIES

Mapping is the digital graphic representation of the facilities and digital interface of the facility attributes with the graphic data.

### Graphic Representation

Graphic representation encompasses the mapping and pictorial communication requirements while maintaining the graphic standards.    The interactive graphics workstation is the tool used for this communication. Behind the scenes are a number of files created to fit your specific standard mapping practices and procedures.    The manual system is maintained with drafting standards or mapping practices. These are composed of symbologies and textual information describing various elements. The CADD system contains cell and symbol libraries for defining these standard symbologies and font libraries for defining standard textual information.

CADD system programs are used to structure standard procedures. These automated procedures force desired input or output results. This is a definite advantage over manual practices which can only provide guidelines to standards. A menu is used for selecting practices, procedures and system functions at the interactive graphic workstation. For a sample menu, see Figure 3.

CADD procedures via menu use the mapping practices to create drawing files. The files use the overlay concept. The files may include, but are not limited to, facility maps, operating maps, construction order drawings, boundary maps and index maps. Figure 4 shows the automation steps.

The following paragraphs list the automation suggestions for some of the desired functions of a water department:

## FACILITY MAPPING RECORDS

- Layer Storage
  easy access to desired land and water facility information

- File Update Protection
  from unauthorized departments or controlled users

- Multipurpose and Multiscale Maps
  district maps
  distribution maps
  systems control maps
  pressure area maps
  valve maps
  cathodic protection maps
  operating maps
  leak survey maps
  index maps
  land and geography maps
  others

- Irregular Maps
  produced across map boundaries and for irregular shapes
  political boundary maps
  engineering, planning and other special study maps
  others

- Reference File Access
  engineering designs
  pending construction
  accounting
  facility details
  others

- Map Management
  easy access to maps and facility details

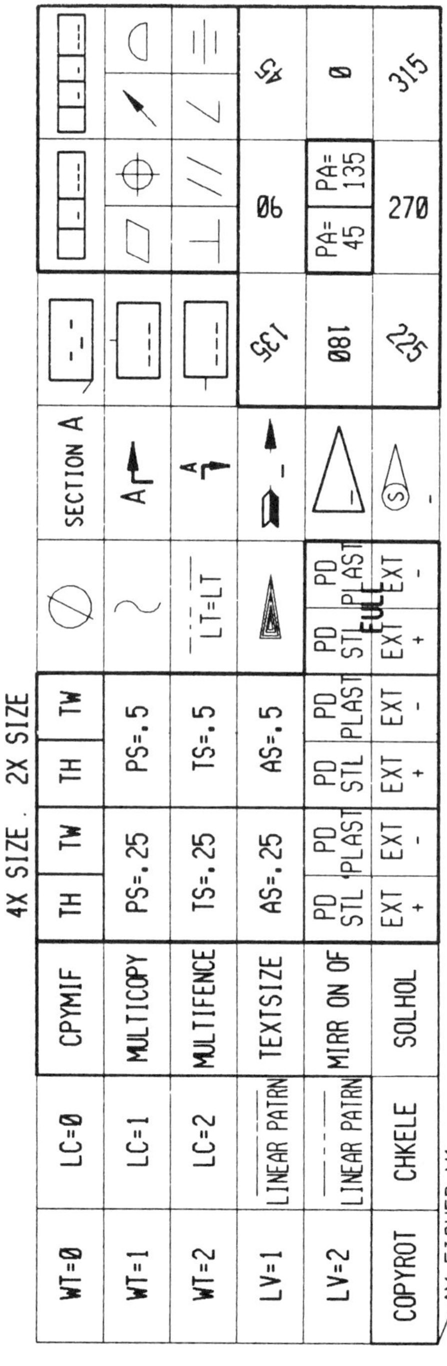


Figure 3. Sample menu.

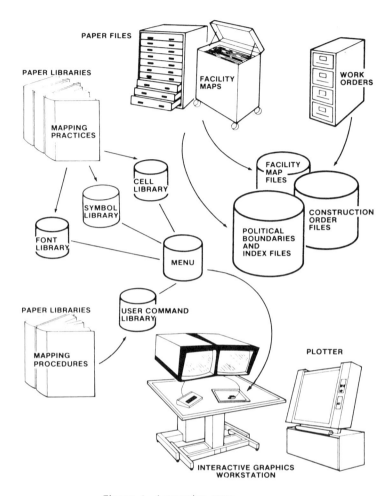

**Figure 4.** Automation steps.

. Complete Coordinate Storage
    continuous grid mapping
    direct merge of geographic located data

# CONSTRUCTION ORDERS

. Work Order Sketches
    selectable geographic areas
    selectable scale and blowups
    selectable facilities and symbology

. Reference File Access
    water facility records

other engineering design work
pending construction
other utilities
land and rights-of-way

. Work Order Status
pending and as constructed symbology
automated graphic status update

. Display, Plots, or Graphic Reports
by work order identification and status

# BOUNDARIES

. Political Boundaries
city, state, township, etc.
tax areas
special districts

. Corporate Boundaries
franchise or service boundaries
districts or divisions
corporate limits
control areas
engineering areas
service planning areas

# WATER MAPPING PRACTICES

. Cell and Symbol Libraries
cell, rapid display, and plot
symbol, expanded for special map generation
standardization of mapping symbology
geography
mains
services
cathodic production equipment
control equipment
other facilities
as constructed facilities
proposed or pending facilities
others
multiscale mapping
multipurpose mapping
collective modification of map symbology standards

. Font Library
standardization of map product lettering

geography attribute lettering
facility attribute lettering
capability for many styles
simple lettering styles for fast display
lettering along curves
collective modification of map lettering standards

# WATER MAPPING PROCEDURES

. Time Reduction for Repetitive Functions
data conversion procedures
data update procedures
standard map production
standard graphic reporting
standard tabular work
calculation procedures
precision placement of facilities
English/metric conversion

. Procedure Control
increase data integrity
tailor procedure control to meet the needs of each department
and function allow selective data usage
establish and maintain facility system model
mains, in line equipment services, customers, etc.

# MENU

. Interactive
instant access to the water procedures and practices libraries
cell commands
symbol commands
font commands
user commands

. Tailored
multimenu capability
designed for each department or discipline as required
conversion
update
engineering design
operations and maintenance
accounting
general access

It is very difficult to justify computer aided maps for the sole reason to draw
and manage the water facility. The maps have many other benefits in the operation
of various government departments. The state, county and the local governments may

jointly fund the effort, and set up the common map standards. The utilities in the area will always be interested in such maps. The accuracy needs are generally more stringent for the government than for the utilities. Sharing the effort will reduce the financial commitment of any one entity. Other maps such as digital terrain models may be effectively used for pipe layout, construction estimates, surface run off predictions and flow control analyses. The water departments would be interested in these.

## NON-GRAPHIC DATA LINKAGE

This involves establishing and maintaining the facility detail description records and the facility system model. The individual records can be created from the inter-active graphics workstation, an alphanumeric terminal, or by tape transfer from exist-ing facilities files. The geographic location of the records (facilities) and the interrelationships or system model are created at the interactive graphics workstation (Figure 5).

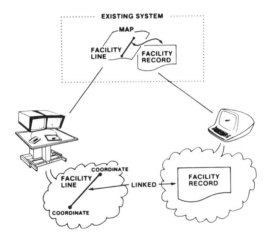

**Figure 5.** Interactive graphics workstation.

## Facility Records

A detail record is created and maintained for each plant item or facility. These records are extremely flexible in providing necessary descriptive items for each facility. Each facility type is developed to fit specific description (attribute) requirements. Code list files are also available to be used for known valid values for specific facility attributes. Code lists might be values such as account codes, equipment size, manufacturer, etc. (Figure 6).

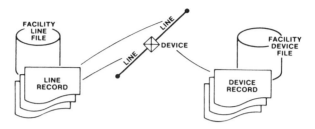

**Figure 6.** Facility descriptive code list.

## Facility Files

Each facility type is maintained in a separate file. Thus, facility line records and facility device records, for example, are maintained in separate detail facility files.

## System Model Relationships

The facility map vividly illustrates the complex relationships of the facility system. The schematic representation is designed to provide visual understanding of the system configuration. This visual schematic representation must be converted to a digital form. Thus, each relationship must be established and maintained to have a digital representation (model) of the facility system. These relationships are defined at the interactive graphics workstation as a function of automated mapping (Figures 7 and 8).

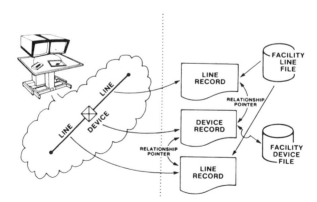

**Figure 7.** Facility map.

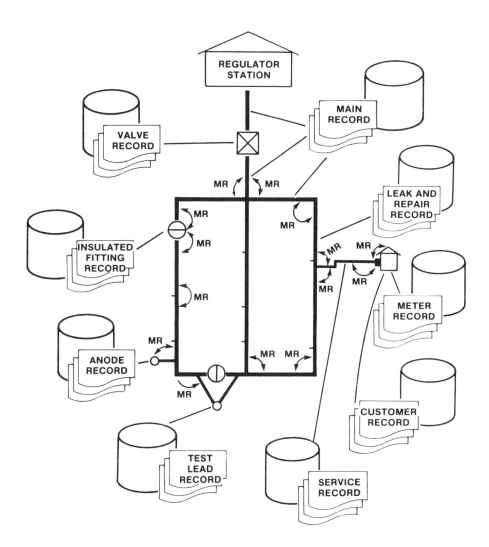

**Figure 8.** System model relationships.

· MAIN SEGMENT FILE

main segments number
material
coating
size-O.D.
wall thickness
length
location measurements
insert description
maximum allowable
   operating pressure
pressure test
date installed
status
manufacturer
account code
work order
tax code
depth
Etc.

· TEST LEADS/POINTS FILE

type
location measurements
work order
date installed
inspection date
inspection voltage
conclusion
frequency of maintenance
trend analysis of readings
existence of electric bond
   to non-gas facilities
Etc.

· ANODES/RECTIFIERS FILE

type (or material)
size
date installed
manufacturer
work order
device number
account code
location measurements
depth
backfill
Etc.

· VALVE FILE

valve number
type
manufacturer
model
location measurements
size
turns to operate
pressure rating
position status
work order
status
inspection/maintenance code
date installed
account code
tax code
valve enclosure data
emergency procedure number
depth
Etc.

· LEAK/REPAIR FILE

location description
survey date
leak detection method
inspection number

· FITTINGS FILE

type
location measurements
manufacturer
work order

general pipeline condition
leak classification
repair date
repair code
follow up inspection date
follow up inspection code
replacement w.o. number
Etc.

date installed
account code
insulation code
Etc.

. SERVICE FILE

building address
location description
size
length
material
status
date installed
tap fitting
tap size
service valve type
service location
cathodic protection/anode
extension deposit number
work order
account code
depth
Etc.

. METER FILE

meter number
location
manufacturer
serial number
load
delivery pressure
number of dials
date purchased
date installed
diaphragm type
last date tested
sample test group code
regulator number
regulator size
regulator make
regulator orifice size
pressure class
relief setting
accounting code
Etc.

. CUSTOMER FILE

address
street name
status
account number
centroid
customer code
Etc.

## SCHEMA (Data Base Structure)

The data base Schema is written to establish the data format for facilities and their interrelation. The schema file defines the complete data base structure: (Figure 9)

. facility files
. field length and configuration (for all facility files)

**The data base schema is designed to be tailored to fit your data base requirements.**

The schema file defines the complete data base structure:

1. Facility files
2. Field length and configuration (for all facility files)
3. Code lists
   - Valid data items (for standardized fields)
4. Facility system model relationships

### FACILITY — FILES

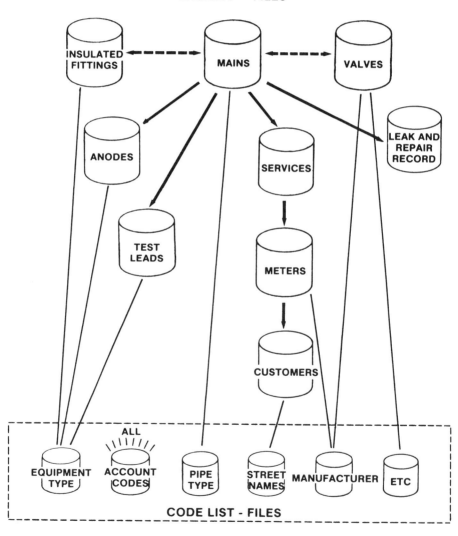

**Figure 9.** Schema (data base structure).

- code lists
  - valid data items (for standardized fields)
- facility system model relationships

## CONCLUSION

Once an interactive graphic system is established, the database of the distribution facilities supports many diverse applications. Work order processing can assist in the maintenance of engineering design and construction status records. Facility maintenance operations and plant accounting can become more efficient through the use of standard and demand report generation. A single master map system is maintained. Various map products are generated, eliminating the need for manually preparing numerous maps of different scales and symbologies.

# Chapter 11

# Power Usage Optimization and Control
# by Computer

Thomas S. Ritter, P.E.
McNamee, Porter and Seeley
Ann Arbor, Michigan

## INTRODUCTION

Clean water, like most benefits of civilization ad technology, uses energy – and costs money. The more sophisticated our methods of treating water and wastewater, the greater our consumption of electrical energy will be. This higher usage of electrical power, and its increased cost per unit, have brought about an increasing effort to control the use of this resource. Like many of the difficult problems of our world, control over electrical power use and distribution can be managed effectively through computers and computer based products.

Let's first consider the characteristics of water and wastewater plants that call for electrical power and control systems. Incoming water or wastewater passes through the process plants, changing continuously as it goes through each stage of treatment. Each stage is progressively larger, resulting in plants spread out over many acres.

Some of the processes can be very energy intensive in one location, like high service pumping or secondary aeration. Other uses of energy are smaller but spread out geographically in equalization, filtration, settling basins and, when we consider the collection and distribution network, in remote pumping locations.

The sophistication of control for each operation depends on the type of operation. Simple pumping operations have pumps that may respond to float switches or pressure switches. Filtering operations require more sophistication whether backwashing is done manually or automatically. In this case, the controller must know the loss-of-head, backwash flowrate, when to turn on or off a surface wash system and the amount of backwashing necessary. Therefore level sensors, flowmeters, timers and control relays must interact to achieve the desired results. But control sophistication does not stop there. If we try to control the pressure in a water district, we need to know the pressure level in various points in the community, the status of the storage tanks and which wells to turn on or off. We now add to our list of control devices telephone lines, remote controls and pressure sensors.

At this point, as control engineers, we would stand back and conceptualize a kind of controller that could provide all sorts of sophisticated control over a process without requiring a huge control panel full of discrete components, any of which could fail and halt the whole operation. Then add to the controller specification the ability to gather and distribute data at some remote operation and communicate with a central control location.

What I am describing is a computer-based controller, one that can perform both the most mundane control as well as the most sophisticated. When it is part of a distributed control scheme, it communicates with other computers over a twisted pair of wires. What we are interested in today is how to use this computerized controller to control the electrical energy we use in our plants. In order to do that, we are going to first look at how a typical utility charges us for the energy we use. Once we understand the utility rate structure we will see how we can minimize what we pay by exploiting the capability of the computer controller to control remote loads economically and provide sophisticated control for an energy intensive process. In the end, we will take a brief look at power factor control.

## ELECTRICAL ENERGY RATES

It is important to keep in mind that electrical energy costs have been increasing at rates of 8 to 10% per year in Michigan and that there is no likelihood that the growth in rates will stop. The water and wastewater industry is a very energy intensive industry. Therefore, you must not only provide excellent results, whether it's a glass of drinking water or staying within your National Pollutant Discharge Elimination System (NPDES) permit; but you must also understand how to use electrical energy efficiently.

Electric utilities have numerous rate structures. Rate structures are established to encourage certain patterns of energy usage that benefit their power system. That is, utilities would normally be faced with a large demand for electricity during the day and very small demand for electricity at night. This would mean that they would have to make a very large investment in generating capacity to meet the daytime demand. This equipment would then be left idle at night. For this reason, utilities have encouraged night time usage by offering reasonable rates for users of electricity at night. For example, the municipal street lighting rate structure generally charges less per kilowatt-hour (kWh) of energy than would a typical, day-and-night domestic energy user.

For large users of electricity, such as water and wastewater treatment plants, the utilities offer rate structures based on the primary voltage, that is, voltages larger than 600 volts and typically around 12,000 or 13,000 volts. These rates structures are advantageous to the utility because the user makes the investment in the transforming equipment. Primary rate structures have billing parameters designed to encourage the user to control the maximum demand for energy, the time-of-day usage and efficiency. Stated another way, there are costs for demand, for energy and for power factor.

## ELECTRIC DEMAND

Electric demand is simply the rate of energy consumed instantaneously, that is, the sum total of all loads in operation at one time.

The primary power rate structure is designed to limit the ratio of maximum demand to minimum demand. For example, a utility would be economically better off to deliver power continuously to a 500 kW load rather than deliver 500 kW for 1 hour and then 1 kW for 23 hours. The reason is not solely because the utility sells more kWh of energy in the first case - which is what the utility wants to do.   It is also because the utility makes an investment in transmission and distribution equipment that is idle in the second case.   In the second case, the utility has 499 kW of capacity that is unused 23 hours a day.   Therefore, by establishing demand charges, the utility is encouraging the user to maximize the efficient use of the utility equipment.

Typical demand charges are $9.50 per kW; this makes up 50% of your utility bill. Based on these figures, a 10% reduction in demand could mean a 5% savings in electrical costs.   As we shall see, savings in electrical demand can occur without a reduction in electrical energy consumption.   If the maximum electrical demand were to occur at noon and the minimum at midnight, simply moving a portion of the noon demand to midnight would result in a reduced electrical bill.

## ELECTRIC ENERGY

The easiest way to think about elecrical energy usage is to think of it as little units of energy called kilowatt-hours (kWh) and for each kWh we use, we pay a price. We are all familiar with it because this is the way we pay for our energy at home. For our treatment plants, this is only one portion of the electrical bill.   Whereas most of us are paying 5 to 9 cents for every kWh we use, water and wastewater treatment plants are paying 3 to 5 cents per kWh.

This range of cost for a kWh is sensitive to the time of day when it is consumed.   If we use a kWh of energy during the day it may cost 4 cents, since this is considered an on-peak time period.   However, if we use a kWh of power in the evening it may cost 2.5 cents, since this is the off-peak time period.   This is the utility's way of informing us that the production of kWh is more expensive during the day and that there is a greater demand for it.

In Michigan, the two largest utilities are Consumers Power and Detroit Edison. Consumers has designated three time  periods: on-peak (from 10:00 a.m. to 5:00 p.m.), intermediate-peak (from 5:00 p.m. to 9:00 p.m.) and off-peak (from 9:00 p.m. to 10:00 a.m.).   These hours are subject to seasonal changes. Energy costs are approximately 3.9 cents, 3.54 cents and 3.24 cents, respectively.   Stated another way, on-peak costs are 22% higher than off-peak costs.

Detroit Edison has two time periods factored into their Primary Pumping Rate structure.   The on-peak hours are from 11:00 a.m. to 7:00 p.m., and the rest of the time is off-peak.   Energy costs are approximately 3.7 cents/kWh and 3.0 cents/kWh, respectively.

Being charged for energy by these rate structures encourages the user to control energy two ways: use only what you need and use it during off-peak time periods.

## POWER FACTOR

The power factor charges in a rate structure have nothing to do with the consumption of electrical energy in kilowatt-hours.   Power factor charges are strictly assessed as a penalty. That is, if you are operating a plant that is not taking kWh effectively from the transmission lines you will be penalized. To understand how this works, we must first understand what power factor means.

Power factor is a measure of how much useable power is transmitted to the user along with unuseable power (called reactive power) that is transmitted back and forth between the utility and the customer.   Often the reactive power is referred to as "imaginary power." Then we refer to the utility as providing real and imaginary power. The concept of imaginary power is at first difficult to understand.

Fundamental to understanding power factor are the two types of electrical elements. First, there are devices that use electrical energy, and second, there are those that store electrical energy. Resistive type devices use electrical energy. An incandescent light bulb is a good example. All of the electrical energy applied to it results in heat and light output. Devices that store electrical energy are capacitors and inductors. Capacitors store energy in an electrical field and inductors store energy in a magnetic field.

Motors, prevalent throughout water and wastewater plants, can be characterized by both resistors and inductors. The resistive portion of the motor uses the energy and performs real work such as lifting water or compressing air. The inductive portion of the motor simply stores and releases energy from and to the utility with each cycle of voltage waveform. Therefore, motors require the utility to provide kilowatthours of energy for consumption and reactive, imaginary power to energize the inductance. Utilities derive no economic benefit from delivering reactive power to their customers. In fact, utilities suffer losses in generating and transmission line capacities because these resources are being utilized for non-revenue, imaginary power delivery. Therefore utilities penalize users that have significant amounts of inductance in theier electrical system. Those with more inductance receive correspondingly larger penalties. Power factor is the measure of this penalty and is measured as follows:

$$\text{Power Factor} = \frac{\text{real power}}{[(\text{real power})2 = (\text{imaginary power})2]\ 1/2}$$

The best power factor is 1.0 and occurs when there are not imaginary power requirements. The worst is 0 and occurs when there is no real power delivered to the user.

Detroit Edison requires the monthly power factor to be above 0.85 to avoid any penalty charges. Consumers Power Company requires the user to stay above 0.80 on a monthly basis to avoid a penalty and will provide the user with a credit for power factors above 0.90. Costs for power factor penalties are usually determined as a percentage of the monthly bill, with an increasing percentage for a poorer power factor.

## TYPICAL ENERGY CONSUMPTION PATTERN

The purpose of building water and wastewater treatment plants is not to use electrical energy efficiently, but instead, to deliver drinking water as needed, or to treat wastewater as it arrives to the plant and deliver an acceptable effluent to the discharge point. When little or no consideration is given to the use of energy in these plants, we can assume that energy usage is strongly correlated to the amount of flow through the plant. A typical 24-hour pattern of flow through a plant is shown in Figure 1.

Generally this flow pattern is used to describe both water and wastewater treatment plant flow. However, water plant flows would normally peak before wastewater plant flows since there are delays in flows because of pipeline and storage tanks. The purpose of this graph is to show the pattern of consumption of water, which usually is at a minimum in the evening hours and begins to increase around 6 a.m. High

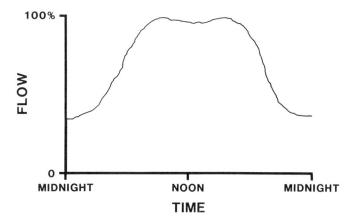

**Figure 1.** Typical water consumption pattern.

usage is maintained throughout the day and is diminished as the evening hours approach.    Note that the area under the curve represents the total quantity of flow delivered to the water system or received at a wastewater treatment plant.    We could also display the same flow pattern in terms of energy usage (Figure 2).

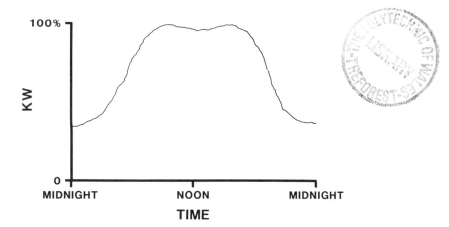

**Figure 2.** Typical energy consumption pattern.

This graph illustrates how increasing flow causes increasing expenditure in energy. For example, when demand for water increases, the output of pumps must increase. Increasing the flow can only be accomplished by increasing horsepower, which results in more consumption of kilowatt-hours.

The graph is scaled 0% to 100% so that it can be applied to any plant.    In other words, a plant with a maximum demand of 10,000 kW could be characterized by this graph simply by multiplying the vertical axis by 10,000.    The area under the curve represents the total amount of energy used in terms of kilowatt-hours.

When we look at the rate structure of the utilities, we realize that users of water are not establishing a consumption pattern that naturally minimizes the costs for electrical energy. The maximum to minimum demand ratio is approximately 2.5:1 and large portions of kilowatt-hours are being consumed during the on-peak hours established by the utilities.

An optimum electrical energy consumption pattern has been overlaid on the natural energy consumption pattern in Figure 3. The optimum energy consumption pattern is a steady demand for energy or a ratio of maximum to minimum demand of 1.0. The area under this curve is equal to the area under the varying demand curve; that is, the total energy in kilowtt-hours is the same for both curves.

Clearly what has to be done to achieve these savings is to shift the demand. This means trying to accomplish certain operations earlier or later than normally would be expected, for example, topping-off an elevated tank before on-peak hours, or postponing filling until after on-peak hours. All decisions are, of course, based on what the maximum demand limit is and whether the process can be interrupted from its normal operation.

## COMPUTER USAGE

Now that we recognize what the electrical rate structure is and the energy consumption pattern that we must deal with, we can begin to work on the task of optimizing the energy usage. Specifically, we want to know how to use a computer to monitor, analyze and control an energy management system.

Two tasks the computer must control to optimize dollars spent on electrical energy are first, shift demand, and second, more work with fewer kilowatt-hours (Figure 3). Shifting demand to get low cost kilowatt-hours implies that there is storage capability within the process. The computerized energy management system must maximize the storage capabilities of the process.

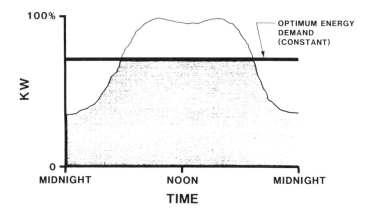

Figure 3.  Typical and optimum energy consumption patterns.

## Demand Control

Demand control at a treatment facility is feasible with a computer based system if we utilize the computer's ability to control at a distance. Since the loads are distributed over many acres we must have an economical means of gathering information about the status of the process and be able to turn it on or off. If we were required to hardwire every data and control point, the cost of wire, conduit and labor could negate any savings in electrical costs. When we have accessed all of the data points and control points within the plan, we can begin to think in terms of energy management. Energy management covers the whole area of demand reduction through time-of-day scheduling and load cycling. Dedicated energy management systems have most commonly been applied to commercial buildings but they do possess features which could be useful in a water or wastewater treatment facility, especially one that employs distributed process control.

In operation, energy management systems are connected into the motor starters for the equipment items to be controlled and monitored by the system. The system monitors plant load continuously and, at some adjustable level of energy consumption, will shut down or inhibit the operation of selected pieces of equipment in a predetermined sequence. Normally heating and ventilating systems are the first to go. For time-of-day scheduling, the system will inhibit certain loads from operating during on-peak hours and enable operation during off-peak hours. There may be some plant loads which could by cycled on and off over the course of the day rather than operated continuously. Any number of such loads could be automatically cycled by the energy management system. A typical example would be to inhibit the filter backwashing operations until off-peak hours. Then during the off-peak hours individual filter backwashing might be alternated (cycled) with another process - for example, heating and ventilating.

An attractive feature of a dedicated energy management system is its data management and report generating capability. The more sophisticated systems have the capability of generating printed reports depicting the status of all monitored equipment, the times during the day that this equipment was started and stopped, the energy consumption trend over a selected time period within the facility, etc. This type of data is useful in optimizing the plant operation to minimize demand charges.

Energy management systems used to limit electrical demand have been used extensively in commercial installations but are not widely used in treatment facilities. There are several reasons for this. First, in applying an energy management system to a continuous, non-interruptible process such as that encountered in a water or wastewater treatment plant, care must be taken in choosing those pieces of equipment to be controlled by the system. Some equipment can be safely shut down without having an adverse effect on the treatment process. However, interrupting or inhibiting the operation of equipment vital to the treatment process can sometimes have disastrous results. In addition, analyses must be performed to determine the impact of an energy management system failure on the treatment process. The system should have a means of being manually bypassed and should shut down safely.

Secondly, there are many less sophisticated products that are competitive with a totally computerized system. Time clocks, for example, are used on nearly all new HVAC systems, photo cells automatically control outdoor lighting and motion detectors are being used on many interior lighting applications. These devices significantly reduce the plant demand, making it difficult for a sophisticated energy management system to be cost competitive on an incremental basis.

Thirdly, simply restructuring plant operations can be done without an in-house computer. Plants that are on demand-type rate structures soon find that the monthly utility bill provides useful information about demand peaks. The utility's meter gathers the peak demand, energy usage and power factor data that is printed on the monthly bill. The user then knows what the maximum demand is and when it occurred. This can be compared to the average demand that can be calculated by dividing the hours in the billing period into the total kilowatt-hours consumed. If the maximum demand is appreciably different from the average demand, the user generally makes changes to perform more operations during off-peak periods.

Finally, demand control systems have not proliferated throughout the market because many treatment facilities are not on demand-type rate structures. Many municipal facilities are fed from local municipal utilities, which typically do not have demand type rate structure. Under energy only rate structures, the user is encouraged to minimize his kilowatt-hour consumption. Therefore, we will now turn our attention to minimizing kilowatt-hours.

## Energy Reductions

Having optimized our maximum to minimum demand ratio, we can begin to work on our second task - minimizing our energy consumption. During the 1970's and 1980's, the municipal market for energy efficient products greatly expanded. To fill this market, manufacturers expanded their product lines to include specialized (and higher priced) products that provided reasonable payback periods through energy savings. These products include high efficiency motors, variable frequency drives and energy savings lamp and ballasts.

Complementing these products, we automated our plants and processes as much as possible with analog instrumentation. In the process we gained tighter control over our chemical usage and our expenditure of labor, and improved the quality of our finished product.

Today's process control computer has given us an opportunity to take another step forward. We used to think of a computer as being a large, bulky piece of electronic gear located in a computer room. The computer and to be housed in a central air-conditioned room with humidity control and a raised floor. If we used the computer for process control, it would run the entire process. If the computer failed, the plant would shut down. Means of bypassing the computer were developed to alleviate this problem, but these increased costs. Distributed process control has changed all of this.

Distributed process control is achieved by using microcomputers distributed around the plant. These computers are specifically designed to work in a nonair-conditioned environment where some electrical noise may be present. Each microcomputer is dedicated to a process or area. The microcomputer controls the process, acquires data about the process and communicates with other computers. The failure of this computer does not lead to the shutdown of the plant but only the suspension of the process. This problem is alleviated with many manufacturers offering redundant back-ups. Hence, reliability of the computer based control system is quite high.

Dedicating a microcomputer to a process provides a tremendous amount of computing power and allows us to exceed the capability of analog instrumentation. This will become clear when we examine an example of improving efficiency. In this case, the computer and software replaced a significant number of comparable hardware devices. If we had to use the actual hardware devices, we would not have been able to economically control the process. Keep this in mind as we look at a typical process.

## AERATION SYSTEM

When we begin to think about increasing efficiency at a wastewater treatment plant, it seems natural that we look seriously at processes that are energy intensive and have been beyond our control with conventional analog controllers.    In an activated sludge wastewater treatment plant, the largest user of electricity is the aeration process.    Low pressure aeration blowers can often account for over 50% of the utility bill.    We next ask ourselves if the aeration process offers any appreciable opportunities for improvements.    The answer is yes, and that further improvement may be possible only with a computer.

Typical aeration systems are either being run manually or by automatically positioning an air inlet valve based on process influent flow.    In either case, the variable that is to be controlled, i.e., dissolved oxygen (D.O.), is imprecisely known.    Air is actually added in excess of the process requirements.    Therefore, if we could measure the D.O. in the tanks continuously, we could more closely control the operation of the blower.    Our analysis has shown that there is a potential for a 7% to 9% reduction in air production in aeration systems in Michigan today.    This corresponds to a 10% to 12% reduction in energy consumption.

Clearly the key to achieving these savings is the ability to reliably measure the D.O. in a hostile environment.    Also, should the measured D.O. signal be lost, the blower and control valves must default to a predetermined mode.    The idea of controlling more closely the amount of air added at aeration is nothing new.    There have been installations where D.O. probes were used to control guide vanes through proportional plus integral plus derivative (P.I.D.) controllers.    Unfortunately the D.O. probes have proven to be unreliable due to fouling and lack of maintenance. Also, connecting a D.O probe directly to the controller, without allowing for the response rate of D.O., leads to control loop instability and recorders painting their charts.

Improvements in D.O. probe design have led to a more reliable measurement system. Probe technology now takes advantage of advances in electronics permitting more components to be installed in the probes, as well as increasing the surface area of that portion of the probe in contact with the media.    These features allow the probes to operate longer without maintenance or replacement.

Even with improvement in the probe, the use of the computer or microcomputer is essential for th reliability of an aeration control system.    The computer may be viewed as a one component device, albeit a complicated one with printed circuits for the central processing unit (CPU), power supply, inputs, outputs and a communication network.    Of course, each printed circuit board has a large array of integrated circuits and discrete components.    The integrated computer system has proven to be much more reliable than the discrete devices that it has replaced.

But how did the computer system replace hardware devices?    Devices were replaced by being simulated in software.    These devices are control relays, timers, rate-of-change limiters, comparators, PID controllers, multipliers, dividers, adders and subtractors, to name a few. With these algorithms, the ability to control the D.O. in the aeration tank is manageable.

Figure 4 shows a control scheme for an aeration system.    Each tank is to have a desired D.O. level, set by HK 1-1 and HK 1-2.    Controllers AIC 1-1 and AIC 1-2 compare the setpoint values to actual D.O. levels and each adjusts an air inlet valve accordingly to bring the D.O. level to the desired value.    The D.O. probe is submerged in the tank and is connected to an amplifier that indicates the D.O. value and outputs a standard 4-20 mA dc signal proportional to D.O. to the computer.

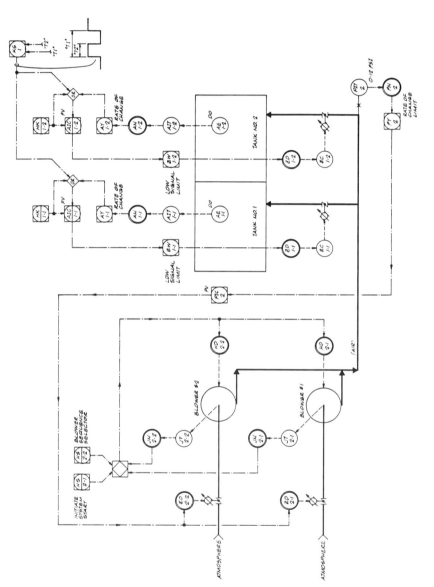

**Figure 4.** Aeration tank air system.

The first task the computer performs on the D.O. signal is to limit its rate-of-change. This is necessary to assure that variations in signal readings caused by poor mixing do not cause excessive valve hunting. The rate-of-change signal conditioning algorithm is an averaging program. D.O. signals are sampled and stored, totaled and divided by the size of the sample. The conditioned signal then is the process variable signal that is input to the osftware PID controller's AIC. The controller outputs a valve position signal to a valve and its position controller through a low signal limiter. The low signal limiter is a comparison algorithm which selects the largest of two values. In this case, if we wanted to inhibit the valve from going less than 10% shut we would select one of the values to be 10%. Then the valve would be limited 10% to 100% open. The purpose in keeping the valve partially open is to assure air delivery to the tank and minimize any potential for the blowers to go into surge.

Time delay can be a serious problem in the measurement of D.O. Delays in response can be 10 to 20 minutes. Therefore, delays in control changes must be built into the control loop. Time lag is accounted for in this control scheme by switching the process variable signal to the controller. That is, the controller process variable signal is either the D.O. measurement or the setpoint value. A time clock signal controls the switching. When the setpoint value is used as the process variable, control action of the controller ceases and the air inlet valve stops moving. When this occurs the control system is actually waiting to see what the impact of the latest valve movement will be. When the clock switches the process variable back to the actual D.O. signal, the controller output begins to change due to the difference in the D.O. in the tank and the setpoint. This causes the valve to adjust in the appropriate direction. The clock signal switches the P.V. signal back to the setpoint value long before the output signal haa a chance to go full scale. Once again, the system waits to see what the impact of the latest valve adjustment is. In this fashion, the valve is iteratively adjusted to the optimum position that will equate the setpoint and process variable. The net result is that the computer has effectively controlled the D.O. in the aeration tank. The air required with fully automated controls will be less than would have been added by manual means where the operator met the NPDES permit by overaerating.

In an aeration system such as this, we must automate the air header pressure control to ensure that the aeration system works properly. If the aeration tanks' air requirements increase, we must have the capability of delivering the additional air automatically.

Before we discuss the pressure control loop we will first look at how a blower works. Figure 5 shows typical curves for pressure and horsepower vs air flow for a blower. These curves show that with an increasing air flow rate we have increasing horsepower requirements. Further, if we regulate the suction pressure of the blower we can regulate the horsepower delivery for a given air flow rate. Since we have the ability to regulate the amount of horsepower we use for delivering air, we need to provide control loops to do it efficiently.

System air requirements are typically established by the air inlet valves to the aeration tanks. In Figure 6, the downstream valve is equivalent to the air inlet valve. With this valve nearly closed, we would have a low air flow rate and high pressure blower output, point A on Figure 6. As the discharge valve opens we progress to point B, the medium air flow rate, low pressure point. If more airflow is required through the discharge valve, the inlet guide vane must be opened. When the guide vane is wide open, the pressure and air flow rate would be as shown for point C.

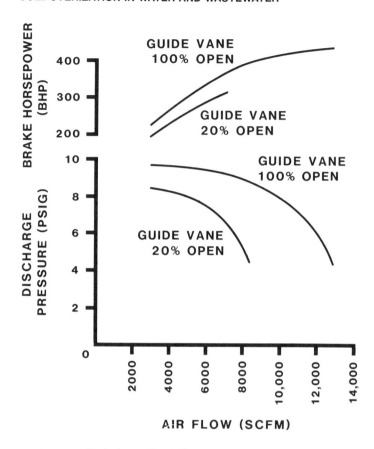

**Figure 5.** Typical centrifugal blower characteristics curves.

The most energy efficient means of delivering the increasing air flow rate in Figure 6 is the path ABC. In practice, this path is rarely followed. The blower discharge header usually has pressure and air delivery requirements which require that a specific minimum air pressure not be exceeded. Taking this into consideration, the engineer sizes the air header piping and selects the appropriate combination of air blowers for parallel operation. The air delivery path is then ABC. This limits the minimum discharge air pressure to 7 psi.

Returning now to our aeration system, Figure 4, we are using a computerized control system to maintain the air header pressure. The computer monitors the discharge pressure and blower horsepower power load, starts and stops the blowers and positions the inlet guide vanes.

Pressure transmitter PIT 2 senses the air pressure in the header and provides an analog input signal to the computer. This signal is conditioned by a rate-of-change algorithm to dampen the random air pressure noise within the system. The conditioned signal then represents the process variable signal which is input to PID controller PIC 2. PIC 2 is used to maintain a pressure of 7 psi or more in the air header by adjusting the blower inlet guide vane. The minimum air header pressure is the setpoint of the controller. Whenever the air header pressure is greater than the

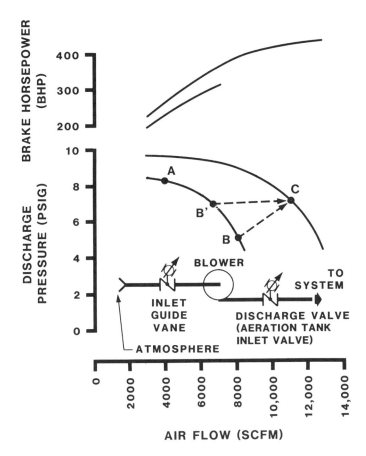

**Figure 6.** Centrifugal blower inlet guide vane control.

setpoint, the output value of the controller decreases by integral action.    If the air header pressure persists above the setpoint value long enough, the output valve will go to its zero value and the inlet guide vane will go to its minimum closed position, usually around 15% to 20%.    Vice versa, if the air header pressure were to decrease below the setpoint value, the controller output valve would increase, causing the guide vane to open and increase pressure.

Start/stop blower control is also handled through the logical operations of the computer.    The operator first selects the blower sequence, through HS 2-2 determining which blower is lead or lag.    The lead blower is to run continuously and is started when the operator initiates the system through HS 2-1.    Assume that blower No. 1 is the lead blower and is placed in operation. Its power delivery is monitored by watts to current transducers JT 2-1.    As air delivery requirements increase and when the horsepower rating of the blower is approached the computer starts the lag blower. Both blowers will operate in parallel until the air delivery requirements of the system have diminished to the point where the computer detects (through horsepower load) that one blower is sufficient to deliver the air.    A well designed computer program for this application will include time delays to inhibit the lag blower from starting until the request to start has persisted for one minute.   The same is true for blower stopping.   These delays assure that field sensor noise does not affect the opera- tion of the blowers.   Also, a time delay should be included to inhibit the blower from restarting after it has been shut down.   Large motors on blowers usually have to cool off for a period of time after running; otherwise the life of the motor is in jeopardy.

The desired operation of this control scheme is characterized in Figure 7.   Curves A and A' represent blower delivery capability with one blower in operation and its guide vane at minimum and maximum position respectively.   Curve B and B' represent blower delivery capability with two blowers operating in parallel with their guide vanes at minimum and maximum position respectively.   Curve C is the minimum pressure requirement of 7 psi and is the setpoint value for controller PIC 2.   Curves D and D' are the minimum and maximum system air requirements respectively.

With the air inlet valves to the aeration tanks at their minimum open operation we expect curve D to represent the system requirements.   If we have one blower in operation we expect to be operating at point E.   The guide vane would be at its minimum position because the air header pressure is greater than the setpoint value of 7 psi.   As the air inlet valves open, we expect air delivery to increase along curve A with increasing air flow and decreasing pressure.   If we were to continue beyond point F, pressure would drop below the setpoint.   When this occurs the computer algorithm for the pressure controller opens the guide vane to increase the pressure.   From point F to point G, the pressure controller algorithm is able to hold the discharge pressure at 7 psi.   At point G the computer recognizes the blower has approached its maximum horsepower rating and starts the lag blower.   Since air requirements in the aeration tanks require only the air flow rate as required at point G, the aeration tank inlet valves begin closing, taking us to point H.   As aeration tank air requirements increase, the inlet valves open and we track along curve B with increasing air flow and decreasing pressure until we arrive at point I. At this point if we continued beyond point I, pressure would drop below our desired minimum.   Therefore, the pressure controller algorithm opens the guide vanes as we continue to point J.   When aeration tank air requirements diminish, the same path is taken in reverse.

From this example we can appreciate that a complicated scheme has been brought under control by a computer.   The software enabled us to replace many hardware devices and we were capable of simulating human behavior.   The system in the end is simpler because is has fewer components and is thus more reliable.

At this point we will now turn our attention to eliminating power factor penalties.

## POWER FACTOR ANALYSIS

In our earlier discussion, we pointed out that there are cost penalties that could be imposed by a utility for having a less than desirable plant power factor.   our goal is to eliminate this penalty and to operate the plant more efficiently.   Installing capacitors at motor loads or utilizing synchronous motors are conventional methods used to eliminate power factor problems.   In many cases, these methods will improve the overall plant power factor above the limit set by the utility.   However, there are loads which have inherently poor power factor which cannot be corrected individually.   Current source, variable frequency drive equipment cannot be individ- ually corrected.   Capacitor equipment must be isolated from these loads to be effective. For these and other types of loads, a central station type of capacitor bank is usually

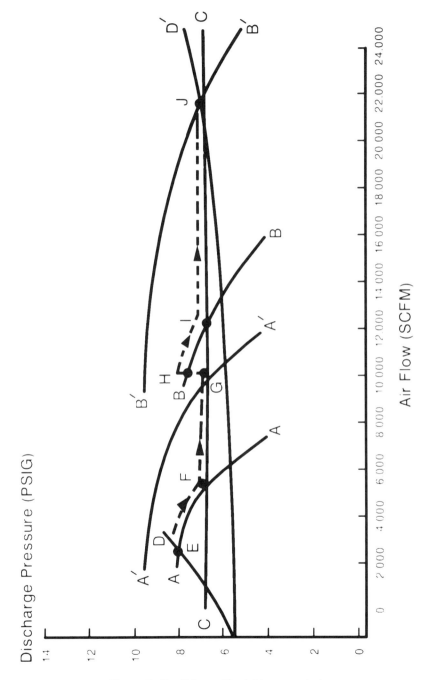

**Figure 7.** Parallel centrifugal blower control.

needed. These capacitor banks need a control system to logically decide when to add or
delete capacitor cells to avoid any power factor penalty. This is a natural problem for
a computer to handle and there are microprocessor-based products available to carry out
the decision making process. The computer used in this application is operated as a
controller with the entire program in read-only-memory. The capacitor bank is made up
of five to ten individually switched capacitors, depending on power factor requirements.
The capacitor bank is attached to the power bus as a load. This connection also provides the voltage input to the microprocessor. Also, a current input is provided for a
current transformer on the main bus. The desired plant power factor is a thumbwheel
setpoint adjustment on the face of the controller. The microprocessor then compares
the setpoint to the actual plant power factor from voltage and current readings. Since
the capacitor cells provide discrete values of capacitance, it is impossible to
correct the plant power factor to the exact setpoint setting. The microprocessor functions
to keep the actual plant power factor above the setpoint but less than 1.0. In this way,
the net result is that at the end of the billing period the power factor has automatically
(by means of a computer) been raised above the penalty point.

Another computer based, though less exotic, solution to the power factor penalty
problem is McNamee, Porter and Seeley's method of analyzing plant power factor. We
began to receive questions from clients concerned about power factor penalties on
their monthly bill. They wanted to know what size capacitor would eliminate the
penalty. We developed our own microcomputer program for analyzing these utility
bills in order to size and recommend capacitors to be added to the electrical system.
In many cases, these clients would then hire an electrical contractor to install the
capacitor or send us a plant one-line and ask for our recommendation on how to apply
the capacitor.

POWER FACTOR ANALYSIS

| | | | | | |
|---|---|---|---|---|---|
| Project: Anytown, USA | | | McNamee, Porter & Seeley | | |
| Proj. Number: | | | Engineers/Architects | | |
| | | | Ann Arbor, Michigan | | |
| Date:  25-Oct-85 10:08 AM | | | (313) 665-6000 | | |
| PF Desired:  85.00% | | | | | |

| | Billing Date | No. Days | kWh | kVARh | PF | Required kVAR |
|---|---|---|---|---|---|---|
| 1 | 7/26/83 | 30 | 72030 | 69045 | 72.19% | 33.9 |
| 2 | 8/24/83 | 31 | 81600 | 84420 | 69.50% | 45.5 |
| 3 | 9/20/83 | 31 | 78510 | 77490 | 71.17% | 38.8 |
| 4 | 10/25/83 | 29 | 74445 | 67890 | 73.98% | 31.3 |
| 5 | 11/23/83 | 31 | 86520 | 80460 | 73.23% | 36.1 |
| 6 | 12/27/83 | 30 | 88410 | 80895 | 73.78% | 36.3 |
| 7 | 1/25/84 | 33 | 102465 | 90975 | 74.78% | 34.7 |
| 8 | 2/24/84 | 30 | 97035 | 88005 | 74.07% | 38.7 |
| 9 | 3/26/84 | 29 | 101985 | 88815 | 75.41% | 36.8 |
| 10 | 4/25/84 | 32 | 204895 | 89385 | 76.11% | 31.7 |
| 11 | 5/24/84 | 29 | 88245 | 74970 | 76.21% | 29.1 |

| | | | | | | |
|---|---|---|---|---|---|---|
| | 335 | 976140 | 892350 | 73.67% | 29.1 | Minimum |
| | | | | | 35.7 | Average |
| | | | | | 45.5 | Largest |
| | | | | | 47.5 | Engineer's Selection |

**Figure 8.**  Analysis form.

Figure 8 is our analysis form. We request the power bills of the last 12 months to be sure we have annual high and low power consumption figures. Our first entry is the power factor desired. This is usually based on what utility is serving the plant. In this example, we want an 85% power factor or greater. Then from the actual utility bills we enter the billing date, number of days in the billing period, the kilowatt-hours and the kilovarhours. The computer then calculates the power factor and the capacitor size in kVAR required to raise the power factor to 85%. The computer also summarizes the columns and determines the minimum, average and largest capacitor needed to correct the power factor. With this data and a review of what size capacitors are readily available on the market the engineer can determine what size capacitor should be recommended. In this example, a 47.5 kVAR capacitor was entered and the computer responded with the lower array of data (Figure 9).

CORRECTED POWER FACTOR REPORT

Using a        47.5 kVAR capacitor

| Billing Date | No. Days | kWh | kVARh | PF |
|---|---|---|---|---|
| 1  7/26/83 | 30 | 72030 | 34845 | 90.02% |
| 2  8/24/83 | 31 | 81600 | 49080 | 85.69% |
| 3  9/20/83 | 31 | 78510 | 42150 | 88.11% |
| 4  10/25/83 | 29 | 74445 | 34830 | 90.58% |
| 5  11/23/83 | 31 | 86520 | 45120 | 88.67% |
| 6  12/27/83 | 30 | 88410 | 46695 | 88.42% |
| 7  1/25/84 | 33 | 102465 | 53355 | 88.70% |
| 8  2/24/84 | 30 | 97035 | 53805 | 87.46% |
| 9  3/26/84 | 29 | 101985 | 55755 | 87.74% |
| 10  4/25/84 | 32 | 104895 | 52905 | 89.29% |
| 11  5/24/84 | 29 | 88245 | 41910 | 90.33% |
| | 335 | 976140 | 510450 | 88.64% |

Figure 9.   kVAR capacitor data.

## CONCLUSION

We have seen how computerization optimizes energy usage in the water and wastewater industry. The impact of energy usage, power demand and power factor on typical electrical energy consumption patterns is critical to our efforts to increase plant efficiency and reduce energy costs. The computer's ability to monitor, control and analyze allows us to revise plant operations to achieve an optimum utility bill.

# CHAPTER 12

## OPERATION AND MAINTENANCE
## USING A COMPUTER IN A SMALL PLANT

William R. Gramlich
City of St. Johns
St. Johns, Michigan

## INTRODUCTION

Operation and maintenance of the St. John's Wastewater Treatment Plant incorporates a wide variety of tasks. Reviewing and analyzing these activities revealed their suitability as computer applications. Purchase of hardware and software was initiated and justified on the premise that overall operation and maintenance efficiency and effectiveness would benefit. During the first year, however, accomplishments were mixed with disappointments. While quantitative assessment is difficult, qualitatively it has been a success. This chapter intends to review the applications and experiences of a small plant during its first year of using a computer.

## SYSTEM

St. Johns Wastewater Treatment Plant provides tertiary wastewater treatment for the city's 7,000 residents. After an $8.3 million modification and addition project, the facility began operation in October 1980. Flow to the facility from 2,491 customers is 1.6 million gallons per day. All processes and equipment are operated and maintained by a staff of six, including the superintendent. The facility is manned 8 hours per day, 365 days per year.

Plant effluent is regulated by a National Pollutant Discharge Elimination System (NPDES) Permit, which establishes discharge limits, monitoring and reporting requirements and other conditions. In addition to monthly reports, development and implementation of a Program for Effective Residuals Management (PERM) and an Industrial Pretreatment Program (IPP) are mandated by our NPDES Permit. These programs include regular periodic reports and updates.

In 1983 it was recognized that the volume and complexity of plant activities were increasing. Further review of these activities determined that some were suitable applications for a computer. Investigation of hardware and software resulted in the purchase of an IBM PC XT (IBM Corp., Boca Raton, FL) and an EPSON FX 100 (Epson Corp., Nagano, Japan) printer in December 1984. Justification for the acquisition was based primarily on three applications: word processing, preventive maintenance and NPDES reporting.

## APPLICATIONS

Word processing may not appear to be a needed application, especially at a small plant; however, the opposite is the case. In our situation, with a limited staff, all clerical duties rest with the superintendent. A substantial portion of his time is allocated to performing routine clerical tasks. Oftentimes, assistance was provided by the city manager's secretary. However, as clerical demands on the plant increased, she was unable to consistently manage her work and assist in the plant's. Therefore, DisplayWrite 2 (IBM Corp., Boca Raton, FL) word processing software was purchased with the computer. By providing this software, it was anticipated that all clerical duties could be more easily managed by the superintendent.

Successful use of word processing requires a clear understanding of its capabilities, along with patience and practice. Although a comprehensive manual is part of the software package and a thorough reading proved helpful, it was confusing, especially to a novice. It contained detailed instructions and plenty of examples; however, using the program and solving problems as they were encountered proved to be a more effective learning method. Spending time actually using the software can be frustrating and demoralizing initially, but, with persistence and patience, it became easier, and, as it became simpler, confidence and speed followed. Frequent discussions with the city manager's secretary, who also has an IBM PC XT and DisplayWrite 2 package, were extremely helpful.

Tasks which were sometimes put off or dreaded due to their clerical burden, such as bid specifications, Annual Operation Report and Annual Budget Report, are now easily handled. Frequently, it becomes a challenge to find a new or better method of solving a particularly difficult or complex task. Large mailings, necessary in the IPP and PERM, are very manageable. Producing multiple original copies for any number of individuals or companies is not difficult. Word processing has given the plant independence from reliance on outside clerical assistance. Our experience has been that, as tasks become easier to complete, projects which have been shelved due to clerical limitations can now be started and completed. Frequently, it has been questioned whether it is economical to have the superintendent performing these tasks. In our situation, the response is yes. Composing and writing can be accomplished at the keyboard with revisions and corrections made prior to printing. In nearly all instances, the finished document has taken less time to complete than it would have without word processing, and it is professional quality. It has completely eliminated reliance on outside clerical assistance and the associated problems of having personnel, in another location and unfamiliar with the operation, performing plant clerical tasks.

Preventive Maintenance (PM) was another application for the computer. Prior to plant start up a PM program was organized and established. Our PM is thorough, but the record keeping and scheduling were manual tasks and very time consuming. Several packaged PM programs were reviewed; however, we found that most of these packages are oriented to a large plant operation. We did not want many of the functions, such as work orders and inventory control, which are included in these programs, nor did we desire to restructure our PM to fit a program. Therefore, we decided to have software custom written. This provided an uninterrupted transition and allowed us to maintain our numbering and identification systems. With the computer identifying equiment, maintaining the histories, and listing scheduled maintenance, the maintenance chief is relieved of most clerical chores. Although he must enter the completed tasks and update the histories regularly, it is a simpler and less time consuming task than it was previously.

A list of the month's scheduled maintenance is printed, including even routine daily tasks. With this checklist the maintenance man or an operator can perform listed PM tasks on the identified equipment and the likelihood of a maintenance activity being overlooked is reduced. It also eliminates reliance on memory, which may fail and cannot be left on the job for someone else to read. Our PM has become more effective and easily managed.

NPDES reporting was the final application considered for computerization. As of this writing, no software to accomplish this task has been purchased. A number of spreadsheet programs and some specialized software packages have been reviewed though. It is anticipated that this application will be running within the next few months.

## EXPERIENCES

Computerizing various plant functions has not progressed as quickly as anticipated. in hindsight, the time required to become familiar with the hardware and software was underestimated. As is the case in most small plants, St. Johns has a limited staff, each operator has a variety of duties, and the superintendent is no exception. Opportunities to interact with the computer were limited; therefore, the learning process toward computer competency was slow. Substantial time is required to become familiar with a program and its capabilities. Also. considerable time is needed to build files and input data.

Although it is evident that the computer can more efficiently handle the information, entering it is a tedious task. In the case of our PM program. a great many man hours were needed to input all the equipment data, histories and maintenance pocedures. This became a very time consuming activity because the personnel were tentative about using the equipment and lacked typing skills. However, this task was a good learning experience.

Although prior training is not necessary, it is certainly helpful. In our case, only the superintendent had any formal training or experience with computers. Operators were interested in the equipment but tentative and cautious about using it. It required considerable persuasion to get them to the keyboard. Oftentimes they became frustrated when the computer failed to reason a problem out or perform in the way they felt it should. However, with continued practice and discipline they have come to realize the computer follows a specific set of instructions and only does what it is instructed to do.

A number of applications not originally considered have been found. These additional uses are the result of simply having the equiment and using imagination. However, to accomplish these tasks software was necessary. It was felt that a good source of information and software vendors would be available in a computer publication; therefore, we subscribed to PC MAGAZINE, which is IBM oriented. Besides technical articles and reviews of software and hardware, numerous wholesale software companies advertise in this publication. We have purchased programs for less than $70.00, and, though they may lack the sophistication of the name brand software, they have proven adequate for our uses and are within our budget. Such software as an electronic spreadsheet and a database have been purchased and are being used. Other more specialized software has also been obtained, such as a sign maker and an electronic flipfile. Since most of these packages are less complex, they are usually easier to understand and use.

## CONCLUSION

Although there may be inherent problems in using a computer in a small wastewater plant, we have found they are not insurmountable. Our experience indicates that with diligence, patience and persistence the difficulties can be overcome. A computer has made our facility more independent, efficient and effective in managing clerical tasks and PM. As additional acitivites are incorporated onto the computer, future results are expected to be positive. We consider our hardware and software investment cost effective and beneficial for the plant and the city.

# CHAPTER 13

## THE REALITIES OF COMPUTERIZING MAINTENANCE ACTIVITIES AT THE DETROIT WASTEWATER PLANT

Harry W. Bierig, Timothy W. Roe, Donald A. Stickel
Detroit Wastewater Plant
Detroit, Michigan

## INTRODUCTION

The Detroit Wastewater Plant serves the city of Detroit and 76 suburban communities. The population served is three million persons. Average flow to the plant is 700 million gallons per day from residential, industrial and commercial sources. Rainfall and snowmelt from much of the service area also is conveyed to the plant, and can raise inflow to the plant to 1.2 billion gallons per day. The plant consists of primary and secondary treatment, with dewatering and incineration of the resulting sludge. The amount of sludge disposed of each day averages 2400 wet tons. The plant occupies 123 acres and is the largest single plant in the country.

## MAJOR TREATMENT ELEMENTS

To get an idea of the maintenance effort involved in keeping the plant functioning in an efficient way, let's consider some of the major treatment elements. There are:

8   Main sewage lift pumps w/associated bar screens & grit chambers

16   Primary settling tanks

5   Intermediate lift pumps

2   Cryogenic oxygen plants

4   Aeration tanks

25   Secondary clarifiers

3   Centrifuges

12   Sludge thickening tanks

28  Vacuum filters

16  Belt filter presses

14  Multiple hearth incinerators

Add to this list chlorinators for disinfection of final effluent; belt conveyor systems; laboratory equipment; chemical addition systems; instrumentation and control systems; ash conveyance systems; sampling systems; process water systems; all related pumps, valves, drains, piping, heating and ventilating systems; electrical distribution system; grounds maintenance and building maintenance. and the magnitude of the maintenance effort required begins to come into focus. Also it must be remembered that equipment in a wastewater plant is subject to constant hard abuse. The plant must operate 24 hours per day, 365 days per year – we never close. Sewage is oily, sticky, abrasive and nonuniform, putting equipment in one of the worst operational environments possible. Total staffing for the plant is more than 1,000 persons, with operations and maintenance personnel always in attendance.

## CORRECTIVE VS PREVENTIVE MAINTENANCE

Maintenance must be thought of as more than just fixing things. The maintenance effort gives the plant capacity; that is, the capacity to operate. The incinerator can't be operated if the conveyor feeding it is not repaired. The final effluent will not meet regulations if the aeration tanks are out of service. Equipment maintenance and operation must go hand in hand, for neither can succeed without the other. In providing the capacity to operate, maintenance performs two basic functions: corrective, or remedial, maintenance, and preventive maintenance. Corrective maintenance is the repair of a piece of equipment that is broken and not available for service. Preventive maintenance is the performing of various tasks to prevent breakdowns of equipment and prolong the operating life of that piece of equipment. These two types of maintenance relate to each other. If there is no preventive maintenance (PM), corrective maintenance tasks are many and much equipment is routinely out of service awaiting repairs. As more and more PM is performed, corrective maintenance tasks decrease and more equipment is in service for longer periods of time. The use of PM will never eliminate the need for corrective maintenance, however. Two examples from the Detroit plant will show the beneficial effect of PM.

Belt filter presses to dewater sludge were put in service in January 1981. After the first year, in which only corrective maintenance was performed, the percentage of belt filter presses available for service was 50%. Since this number of filter presses was inadequate to meet the needs of the plant, a manual PM program was instituted. As a result of this PM program 68% of the belt filter presses were available for service in fiscal 84-85.

The plant multiple hearth incinerators, where the sludge cake produced by the filter presses and vacuum filters is burned to produce a sterile ash, are another example of the benefits of a PM program. For years incinerator maintenance was done on the basis of fixing only what was broken and keeping the incinerator out of service at that moment. As soon as the broken element was repaired the incinerator was returned to service to await the inevitable breakdown of the next element. This approach to incinerator maintenance yielded 40% of the incinerators available for service over the course of a year. In 1983 a new program was begun. Two incinerators, chosen because they were known to be in the worst shape, were taken out of service and thoroughly inspected inside and out and all deficiencies whether known before the inspection or found during the inspection were repaired. These two incinerators were then returned to service and the next two worst incinerators were taken out of service and the process repeated. After three years of this program, 72% of the incinerators are available for service on an annual average. As can be seen by these two examples, PM pays off in keeping equipment in service, and equipment in service means a more consistent effluent in compliance with regulatory limitations and reduced operating costs due to increased equipment efficiency.

Thus, while it can be seen that PM has benefits, it is not possible to perform PM on a manual basis for a plant as large as the Detroit plant. Preventive maintenance activities for only the mechanical and electrical trades were estimated to include approximately 34,200 separate work activities for the total plant. The addition of instrumentation PM is expected to nearly double this number. Instrumentation PM is presently being defined and is expected to take three years to fully implement through the computerized system. To schedule, issue and keep records of work performed is well beyond the capabilities of any manual system for this number of PM activities. An automated maintenance management system was needed. Therefore, in 1982 it was decided to perform both PM and corrective maintenance activities utilizing a computer in order to keep more equipment in service. A consultant to assist in this effort was selected and work got underway.

Before describing the computer system hardware, software and configuration, let us take a look at the various elements needed for a successful maintenance program.

## COMMUNICATION

There must be a way to let maintenance know of a breakdown of equipment and the need for corrective maintenance. This is called the maintenance order, or work order. There must also be a way to alert maintenance to the schedule for performing PM. There must be a way for the maintenance supervisor to schedule his/her work, both corrective and PM, to assure that all work is performed in a timely manner. An element of this scheduling is to set priorities for the various levels of corrective maintenance work. Then the maintenance organization must be such that the work actually gets done. There must be a feedback loop to assure that all concerned are aware that the work is done and the equipment can be returned to service. Records of labor and material expended must be kept. All of this must be done in a way that does not bury personnel under a mountain of paperwork and forms.

In addition to these maintenance order procedures there is also log book maintenance. The log book is used for two basic purposes. First, it reduces the amount of paperwork, as a notation in a log book does not generate the volume of paper that a maintenance order does. Second, it is a rapid way to get small jobs accomplished. To qualify for maintenance log book work the job must be small (less than two hours), be able to be done by one person and use only parts and supplies that are on hand.

## ORGANIZATION

Maintenance is performed around the clock by six maintenance crews, but the bulk of maintenance activities are performed by the day shift crew, which is by far the largest crew. Specific PM crews were designated to assure that PM is given top priority and is effective. These crews perform only PM work assignments. In the event all PM assignments are completed this crew also performs log book work. These PM crews are headed by a subforeman who reports directly to the area senior foreman. Foremen under the area senior foreman perform all corrective and some log book maintenance work for the area except where assistance is required from the central shop. This day shift crew is broken down under senior maintenance foremen, each of whom is responsible for one of the following specific areas: planning, dewatering, incineration, liquids, central shop and facilities maintenance. Basic assignments for these areas follow.

> **Planning.** All maintenance orders are passed through the planning group, which is a multidisciplined group with experience in all plant treatment areas. Higher priority jobs and those that require only minimal parts and/or coordination are sent on to the respective area senior foreman. On the remaining jobs, the planner makes sure that all parts are available and all necessary coordination is accomplished so that the jobs, when scheduled, may be performed in an efficient and expeditious manner.

> **Dewatering.** The dewatering group is charged with all corrective and preventive maintenance in vacuum filters, belt filter presses and conveyors. The group consists of foremen and multidisciplinary crews assigned to each treatment area.

> **Incineration.** The incineration group, as the name implies, is charged with all work on the incinerators and ash conveyance systems. The group consists again of foremen and multidisciplinary crews.

> **Liquids.** The liquids group is charged with corrective and preventive maintenance on the primary settling tanks, aeration tanks, final clarifiers, thickening tanks, chlorination, chemical addition and sampling systems. The group again consists of foremen and multidisciplinary crews.

> **Central Shop.** The central shop is a multidisciplinary group which is charged with providing assistance as required and requested by the area senior foreman. The shop group also performs many of the large construction type work projects as defined by the planning group.

> **Facilities Maintenance.** The facilities maintenance group is charged with the upkeep and custodial care of all plant buildings all work involving heating and ventilation and maintenance of all vehicles and safety equipment.

The total number of maintenance personnel described above is budgeted at 341 persons. In addition, the maintenance group has internal clerical, purchasing and engineering support.

The maintenance organization as described did not always exist in this configuration of assignments and duties. The organization has evolved in this way as a result of determinations and observations made as the computerized system was developed and implemented.

With this organization in place, it is necessary to set priorities on the maintenance work to be done.   The priorities listed in the Water Pollution Control Federation Manual of Practice were determined to meet the needs of the plant.   The priorities are:

Priority    1 - Emergency
            2 - Urgent
            3 - Important
            4 - Routine
            5 - Contingency

These priorities are used for corrective maintenance work and for log book work.

With the organization described and such tools as prioritized maintenance orders, preventive maintenance and log book maintenance, the maintenance crews work on items in the following sequence:

Senior foremen for dewatering, incineration, liquids and facilities maintenance:

1. Perform Priority 1 work
2. Perform PM assignments
3. Perform Priority 2-5 work
4. Perform log book maintenance

All off shift crews:

1. Perform Priority 1 & 2 work
2. Perform log book maintenance and work assigned by the area senior foremen.

## SPARE PARTS

To assure that maintenance activities are carried out as quickly as possible and to be able to know when to schedule specific activities, it is necessary to know what spare parts are available for performing a job and what spare parts must be ordered. This information has been tied into the computerized system and is available to the planning group.   Parts are identified by what are called EMICS numbers.   These numbers identify specific pieces of equipment and related parts.   With a specific number for each part, it is relatively simple to use the computer to know exactly what parts are on hand.   Items are entered as received by the stockroom and deleted when taken out and used to complete maintenance orders.

Having frequently needed spare parts in the most convenient locations was determined to be necessary and required the development of a satellite stores procedure. Specific entries and withdrawals from the satellite stores stock are carefully monitored, recorded and tracked in the same manner as the main stockroom.

## TRAINING

Training to prepare personnel for use of the computerized system has been varied and has included all plant personnel, not just maintenance personnel.   The initial training was on the computerization of corrective maintenance orders which was the first part of the system actually implemented.    This training included the clerical personnel who received and entered the telephone requests for maintenance orders into the system as well as the persons who would be calling in the maintenance orders and needed to be trained in what information was required to initiate a maintenance order.    Training consisted of classroom and hands on training, and involved considerable follow-up until personnel grew accustomed to the procedure and the inevitable "bugs" were worked out of the system.

Other training concentrated on maintenance supervisory personnel to assure that they would understand the total computerized system and be fully aware of the inputs needed to make the system work and of the results and reports they could expect to receive from it to make their jobs more efficient.    Training of this nature was given throughout the computerization project and included most plant supervisory personnel in addition to maintenance supervisory personnel.

Preventive maintenance was first initiated for the primary settling tanks.    Prior to starting this work, several meetings were held with the maintenance crew that would be assigned to this PM work to familiarize them with the work assignments and forms involved and to impress upon them the importance of the PM work that was about to begin.

The most difficult training task was the initiation of the log book maintenance concept.    The major difficulty involved the interaction of maintenance and operations personnel.    This interaction was to be accomplished by the joint entering of the data in the log book to request maintenance work and to report the completion of the various work items.    Complications came from misunderstanding the original form and from the past practice that "for a little item like that we just asked, we never had to write anything."    These difficulties were overcome through repeated training, simplification of the log book form and emphasis on the fact that there was a benefit to having a written record.

## MANAGEMENT SUPPORT

As with all things that are a change from the accepted routine, management support is an essential part of making personnel accept and adapt to different procedures.    The first essential for management support is management information.    For that reason, review meetings were held periodically with the consultant and top department personnel to review progress, identify delays, problem areas and determine future actions. Meetings were also held at the plant level to review, evaluate and agree upon the same issues.

## HARDWARE

The actual computer system purchased to perform the maintenance management activities is a Burroughs B-1955 with peripherals as follows:

1    Burroughs B-1955 computer with a 2 megabyte operating system

2   . dual disk drives with a total memory of 500 megabytes

1   tape drive for loading and dumping information

6   printers

23  terminals

The computerized system is always operating and offers on-line real time access 24 hours a day.   The only exception to this is stocked parts information, which is fed to the system once a day from a separate memory system running on a large computer in downtown Detroit.

The planning section, including the PM group, performs the majority of the work on the system and has ten terminals and four printers available for their use in generating PM and corrective maintenance orders and obtaining parts.

The clerical section has four terminals used in entering the workmen's daily time card data and for tracking purchase requisitions.

The maintenance engineering section has two terminals used in maintaining equipment data and failure analysis.

Three terminals are used by the engineer in charge of the maintenance group, the maintenance superintendent, and the senior foremen to monitor and track maintenance orders and for management information.

Remote terminals with printers are also available in the stockroom for parts requests and in the operations office for entry of off-shift work orders and obtaining information on the status of current work orders.

## SOFTWARE

The software package purchased was a fleet maintenance package designed for a smaller plant with a central maintenance staff. It was selected basically because it was Burroughs compatible, recognizing that modification would be needed to apply it to the Detroit plant.   As development and implementation progressed, the number of modifications was found to be much greater than originally anticipated.   Major modifications include:

   .   Customizing screens and reports to reflect plant needs and to present information in its most useful form.

   .   Revising the PM part of the program to schedule and track activities on a treatment area basis and a set time basis, and to improve the ease of scheduling PM activities.

## PHASED IMPLEMENTATION

For an undertaking the size of computerizing maintenance activities it was essential to phase implementation. This was partly due to the need to experiment to see how things worked and partly due to the impossibility of doing it all at once.

## Corrective Maintenance

The first use of the computerized system was in the generation and tracking of all corrective maintenance orders.    Use of the computer gave a total picture of maintenance activity that was unavailable before.    It also highlighted the amount of work that became backlogged.    The backlog resulted from three primary causes:

1.   Awaiting necessary parts

2.   Awaiting custody of equipment

3.   Awaiting sufficient personnel to perform the work

The backlog is reported on the Age Analysis Report, which includes both corrective and PM work. This report lists all work orders in backlog by process area and by age, beginning with the oldest work orders first.    The Age Analysis Report also includes a summary sheet which breaks down each process area by quantity of work orders and the estimated hours of labor.    This report, in indicating the current backlog in quantity of work orders, is used for charting trends in the number of outstanding work orders. For the last six months (August 85 - January 86), the backlog has varied in the range of 2000 to 2400 work orders on each reporting period. Through January 18, 1986 the backlog had been reduced to an all time low of 1,927 work orders. It should be noted that a total of approximately 11,000 work orders were generated during 1985.

## Preventive Maintenance

Preventive maintenance was first done for the primary tanks and has worked well to date.    The crew originally assigned to the primary tank PM is now also assigned to perform the pump station and rack and grit PM.    An initial finding is that PM tasks are taking longer than anticipated. This is believed to be due in part to the amount of additional work found as equipment receives PM for the first time.    Subsequent rounds of performing the same PM activities are expected to show a reduction in time required to perform each task and a subsequent reduction in total overtime and backlog.    The entire plant is scheduled to be completely on PM by March 1, 1986.    The number of personnel assigned to PM for the liquids portion of the plant is 14 including supervisors.    An additional 16 personnel including supervisors are anticipated for the solids portion of the plant, for a total of 30 personnel to perform PM.

## PM Crews

The concept of assigning specific crews to do the PM has worked well.    The present cew is proud of their designation as "the PM crew" and little backlog of PM work assignments exists.

## Log Books

The log books, after the initial difficulties encountered, are working well.    These log books succeed in keeping paperwork through the computer to a minimum, yet provide a written record of maintenance work for analysis of equipment failure and problems. Also, the second main purpose of getting small jobs done quickly has been accomplished.

## Spare Parts and Satellite Stores

The identification and marking of all spare parts is an ongoing process expected to be completed in spring, 1986. The satellite stores concept is to be tried on a prototype basis in the main pump station and rack and grit areas. As a part of the satellite stores concept, a tool crib is also being established. The purpose of the tool crib is to have accountability of tools and also to have tools available when needed.

## Work Performance

Thus far the computerized system has been beneficial to work performance. The records of jobs done, materials used and labor hours expended, as obtained from information on the completed maintenance order form and entered into the computer, provide senior foreman and management insights into the performance of work. The computer provides a total of 40 reports. Some of the more widely used are:

Maintenance Orders:
- Status information
- Parts information
- Labor information

Equipment:
- Listing of equipment by plant areas
- Nameplate data, manufacturer data
- Maintenance history (labor and parts costs, work performed)
- Failure analysis (types of failure)
- Parts lists
- Parts on hand

Workmen:
- Table of organization
- Attendance
- Current status

Inventory:
- Vendor information
- Purchase requisition tracking

Budget:
- Labor costs per treatment area
- Material costs per treatment area

Management:
- Maintaining order status
- Workmen status
- Capacity planning

These reports provide information to maintenance supervisors in many ways. The major ones are listed below.

The Maintenance Order Status Report by Senior Foreman. This gives the maintenance order status (active, unissued, waiting for parts, planning, major shutdown), the foreman assigned and the maintenance order priority.

The Age Analysis Report. This report breaks down all active maintenance orders by foreman, priority and age (under one month, one-two months, two-three months, over three months old).

The Completed Maintenance Order Report has work performed descriptions, estimated hours vs actual hours and estimated material cost vs actual material cost.

The Capacity Planning Report shows weekly hours available for a given foreman's group after subtracting average time off, average time spent on Priority 1 and 2 maintenance orders, estimated PM hours and average time spent on log book work. It also shows the estimated backlog of work by hours and weeks, and average overtime hours per week.

This list of reports does not include a management level report where one can see at a glance the overall trends in maintenance performance. That deficiency is presently being corrected. Also, the reports are quite voluminous. Exception reports to highlight areas where management attention and/or action are needed must be developed.

These reports do, however, give accountability to the plant's maintenance effort. This is the first time that true accountability of maintenance activities and actions has been available.

## People vs Computers

The maintenance personnel have adapted well to the use of computers. With a minimum of training and much trial and error effort, personnel are able to readily interface with the computerized system. Many programming changes and modifications have resulted from these initial uses including ways to reduce the size of printouts and ways to more easily access and manipulate information. The easy adaptation of personnel to using the computer may in part be a result of finally having a system to assist in planning, organizing and performing work

In using the computer, we have found various advantages and disadvantages. Advantages of the computerized system:

1. With terminal access, the current status of a work order as well as the time charged to date and the person who originated the work order is readily accessible.

2. With terminal access, the outstanding work orders on any piece of equipment in the plant can be viewed.

3. With terminal access, any employee's time for a particular day can be viewed on the screen.

4. Information and summaries can be reviewed to determine trends, problem areas and discrepancies in the accomplishment of the various maintenanc tasks.

From these reports, corrective action or problem areas can be identified and action taken.

Disadvantages of the computerized system:

1. The major problem originally stemmed from the fact that any change in a system is resisted by those who are to make the system work. Progress has been made over the past six months in reinforcing the benefits of the system.

2. Duplicate work orders are sometimes generated when the work orders are called in to the clerks. Most of these are screened out by the planning group.

3. At present, the system will not accept a maintenance supervisor's time as a craft. This shows up as a minor problem on reports that indicate the work orders as finished but with no time charged to do the work.

4. Reports on many occasions are not accurate or appear not to be accurate. Even though it is time consuming to document and investigate the apparent erorr, it becomes necessary in order to determine if the fault lies in the system , or with the maintenance person or supervisor.

5. One of the most frustrating occurrences is that an error in the system will make a report incorrect. This sometimes leads to the perception that the system is not of any use since if one or two errors show up, it leads people to believe that they cannot rely on the information.

Assurances that these problems are dealt with as soon as possible has maintained credibility of the system at a high level for most people.

## Computerized Procedures

The actual procedures for generating, tracking and closing a maintenance order are:

The maintenance order (M.O.) request is called in and put on a screen by a maintenance clerk. Craft information , estimated time , parts information and special instructions are then added on additional screens by the maintenance planner. Hard copies of Priority 1 and 2 maintenance orders are then immediately generated by the planner and given to the senior foreman for distribution to the appropriate foreman and crew . Other priority M.O.s are held by the planner until the necessary parts are available and any necessary coordination has been planned. A report listing these "ready to go" M.O.s is given to the senior foreman, who decides, with input from operations , which should be scheduled. The senior foreman then notifies the planner and the planner generates the hard copies of the M.O.'s for distribution. The maintenance orders are then tracked by use of either screens or reports until the work is complete. The time spent on the M.O. is recorded by entry of the daily time cards by the maintenance clerical staff, and the parts expended are recorded by the entering of stock transactions into the inventory computer system, which then transfers the information to the maintenance computer system. The record is then complete and is moved to the history file.

Actual procedures for developing scheduling, and issuing and closing a PM work assignment are:

First, the equipment is separated by area and senior foreman responsibility. All PM activities for the equipment, including work description, crafts necessary, crew size, estimated time, frequency and materials necessary are entered on screens and scheduled, with the most even work distribution possible, on a yearly schedule basis. Management can then get a report of the PM activities due for a given week. Hard copies of the PM maintenance orders are then generated and distributed to the senior foremen. Tracking, labor hours and expended materials are recorded by the same procedures as the corrective M.O.'s described above.

These procedures have been worked out, modified and debugged as we have implemented the system. They now work well.

## Improper Use

So far there have been no apparent efforts to crash the computer or engage in any but intended uses. However, to discourage improper use the following procecures are observed.

Very few individuals are allowed access to the main computer room, and then only for specific purposes.

Hardware, including terminals and printers, are located in secure, well staffed areas.

Ability to add, delete, generate reports and manipulate data is governed by different levels of passwords, which are changed periodically and are available only to select personnel.

## Backup Procedure

To avoid losing data in the event of hardware failure, transactions against the database are dumped by planners or maintenance engineers to tape once a day, and the database itself is dumped to tapes once a week. Tapes are also made and stored in a remote location once a month. If problems do occur, the database can be reloaded to the machine and the system brought back up.

## Role of the Consultant

The consultant peformed work in the following areas:

Recommended organizational changes and set up a manual system for a prototype area as the basis for the automated PM system

Identified equipment and tagged equipment with EMICS numbers

Identified parts lists and manufacturer's manuals

Collected manufacturers' information on equipment, including recommended PM activities

Procured computer hardware

Selected and procured software

Modified software

Trained personnel in use of the prototype and automated systems

## Role of the City

The city's part in computerizing the maintenance procedures is as follows:

Finalization of equipment identification and tagging

Finalization of collection of manufacturers' manuals

Assistance in software selection and in definition of software modifications

Finalization of software modifications

Finalization of PM activities

Ongoing support of the computer system

Ongoing training in use of the system

Appropriate changes in maintenance organization and institution of new and/or revised procedures and policies to prepare for implementation of the computerized system

Adjustments to organization and policies and procedures as became necessary during implementation

Commitment to the computerizing system in its development, its implementation and in all necessary follow-up actions

## Hardware and Software Selection

The first step was to select the hardware, due to long lead times for its delivery. Then software compatible with the hardware was researched to find existing programs most in keeping with Detroit's maintenance desires. Then the software was modified to suit the Detroit situation. These modifications included both planned and unplanned changes. The unplanned changes were those discoverd during various stages of implementing and developing the system. In hindsight, the exact opposite procedure would, no doubt, have been better. That is, to first define as precisely as possible what was desired from the computerized system, then to procure existing software most in keeping with the defined needs, and finally to define and procure the hardware most appropriate for the software. Since a computer can handle large amounts of data and produce copious volumes of reports, it must always be kept in mind that the primary objective of the system is to keep equipment operational and not to generate paper.

## CONCLUSION

The point of the entire process of attempting to computerize maintenance activities is that it can be done. Doing it does require considerable commitment, the willingness to make necessary changes, and a lot of hard work by all involved in the project.

The benefits include that personnel can be trained to use the system and can be shown that it helps them perform their work. Computerizing maintenance activities improves record keeping, scheduling and the planning of work. More importantly, through the scheduled performance of regular preventive maintenance, it keeps more equipment in reliable operation for longer periods of time. This is a benefit to the maintenance personnel and the operations personnel. The ultimate beneficiaries are those who pay the rates for the treatment of their wastewater. Undoubtedly, as we fully implement the system, additional benefits will be realized.

# INDEX